VIKING ORBITER VIEWS OF MARS

NASA SP-441

VIKING ORBITER VIEWS OF MARS

BY THE VIKING ORBITER IMAGING TEAM

M. H. Carr
W. A. Baum
K. R. Blasius
G. A. Briggs
J. A. Cutts
T. C. Duxbury
R. Greeley
J. Guest
H. Masursky
B. A. Smith
L. A. Soderblom
J. Veverka
J. B. Wellman

Cary R. Spitzer, Editor

NASA Scientific and Technical Information Branch   1980
National Aeronautics and Space Administration
Washington, DC

For sale by the Superintendent of Documents,
U.S. Government Printing Office, Washington, D.C. 20402

Library of Congress Card Number 80-600167

FOREWORD

THE VIKING plan to explore Mars was not simple. Two spacecraft were dispatched at different times in the launch window, and put into different orbits about the planet. After they scouted landing sites, some of which were rejected as too risky, two soft-landers were detached and both—incredibly—managed safe entry and set about making detailed surface measurements and returning dramatic, close-in photographs of another world. Meanwhile the two Orbiters continued to encircle Mars, sending photographic coverage of almost the entire planet. In concept, Viking was ambitious to the edge of audacity.

It was amazingly successful. The Landers revealed a desolate, rocky world, provided fascinating chemical information about the surface, and served as weather stations emplaced tens of millions of miles from Earth. The Orbiters have given us more than 46 000 images so far of the planetary surface, in all seasons, lighting, and weather. In a way Viking still continues even though the major mission has ended, for one of the radioisotope-powered Landers still checks in, and at this writing one Orbiter still has enough attitude control gas on board to continue working.

This volume presents a selection from the orbital images provided by one of the longest-running successes in the history of space exploration. They show Mars as an extremely diverse planet. As you study them, it is difficult to avoid the conclusion that, though Viking contributed immeasurably to breaking the code of the Martian enigma, we do not yet confidently understand its dramatic and turbulent past.

May 9, 1980   Thomas A. Mutch, Associate Administrator
                                 Office of Space Science

ACKNOWLEDGMENTS

THE ENORMOUS SUCCESS of the Viking Project was the result of the combined efforts of more than 10 000 people. To name even that group of people who contributed directly to the success of the Viking Orbiter Imaging Experiment would be an almost impossible task. Three names, however, stand out. In the early days of the Viking Project, the existence of the Orbiter Imaging Experiment was in constant jeopardy. James S. Martin Jr., Project Manager, A. Thomas Young, Primary Mission Director, and Dr. Conway W. Snyder, Orbiter Scientist, were unswerving in their support of a high quality orbiter camera. The pictures in this book are, in large part, the result of that loyal support.

Special mention should also be made of the Space Photography Section of the Jet Propulsion Laboratory, which was responsible for overseeing the design and fabrication of the cameras, and the Mission Test Imaging System personnel of the Jet Propulsion Laboratory, who did most of the picture reconstruction and processing.

CONTENTS

Foreword	v
Acknowledgments	vi
Introduction	1
The Viking Mission	3
Earth and Mars: A Comparison	11
The Great Equatorial Canyons	17
Channels	31
Volcanic Features	47
Deformational Features	63
Craters	73
Variable Features	85
Martian Moons	95
Surface Processes	107
Polar Regions	125
The Atmosphere	139
The Viking Landing Sites	161
Global Color	169
Appendix I. Glossary	173
Appendix II. The Viking Orbiter Imaging System	177
Appendix III. Other Sources of Viking Data	180
Appendix IV. Project Viking Management Personnel	182

For the stereo images, a collapsible viewer is included on the inside back cover of this book.

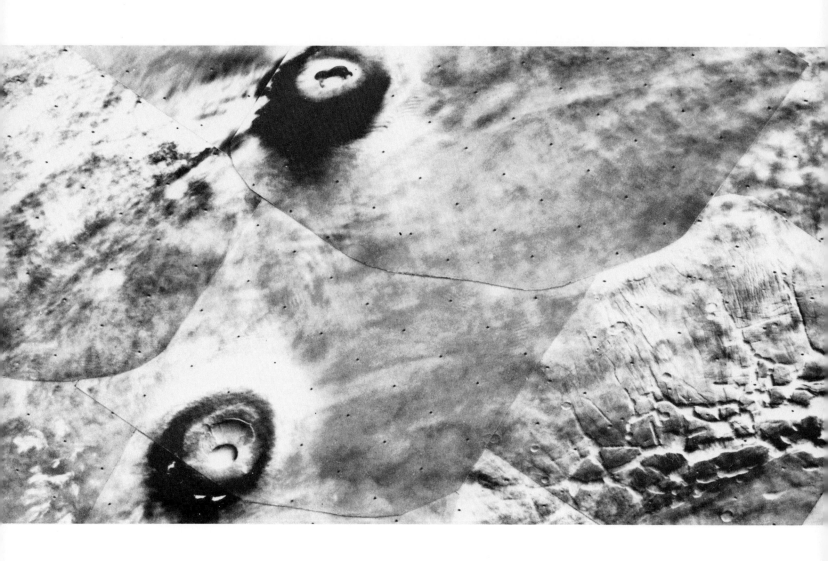

INTRODUCTION

MARS has long had a special fascination for man. As early as the 17th Century, it was recognized that Mars had polar caps and rotated once every 24 hours, much like the Earth. Perception of the planet as Earth-like was subsequently reinforced by observations of white patches that were interpreted as clouds. Late in the 19th Century interest was enormously heightened by reports of canals on the surface. This led to speculation that life might thrive there under climatic conditions similar to those on the Earth and that the canals might be the product of an advanced civilization.

Fact has proved to be nearly as bizarre as fiction. Although the canals were an illusion, and the possibility of life now seems less likely, the planet retains it fascination.

Mars is a geologist's paradise. Many features familiar on Earth are displayed on a vast scale made more awesome by the planet's modest size. Great canyons are incised into the surface, huge dry river beds attest to past floods, volcanoes tower to heights almost three times that of Mt. Everest, and vast seas of sand surround the poles. Global dust storms regularly cover the entire planet. At the poles, caps of carbon dioxide advance and retreat with the seasons.

This book incorporates images acquired by the Viking orbiters, beginning in 1976. The pictures here represent only a small fraction of the many thousands taken, and were chosen to illustrate the diverse geology of Mars, and its atmospheric phenomena. We hope they will also arouse the same wonder and excitement that we experienced on first seeing them.

THE VIKING MISSION

PROJECT VIKING was begun by the National Aeronautics and Space Administration on November 15, 1968. The main objectives of the project were to achieve a soft landing on the surface of Mars and to relay scientific data back to Earth. Scientific goals for the mission were established in response to recommendations of the Space Science Board, a NASA advisory panel. Foremost among their recommendations was that the first Mars lander mission should emphasize life detection experiments to answer the question about the possibility of life on the planet. In consideration of these goals and the everpresent constraints of funds and spacecraft capability, the following investigations and associated instruments were established:

Investigations	*Instruments*
Orbiter imaging	Two vidicon cameras
Water vapor mapping	Infrared spectrometer
Thermal mapping	Infrared radiometer
Entry science	
Ionospheric properties	Retarding potential analyzer
Atmospheric composition	Mass spectrometer
Atmospheric structure	Pressure, temperature and acceleration sensors
Lander imaging	Two facsimile cameras
Biological analyses	Three separate experiments, gas exchange, labelled release, and pyrolytic release, were included to test different biologic models.
Metabolism	
Growth	
Photosynthesis	
Molecular analysis	Gas chromatograph mass spectrometer
Organic compounds	
Atmospheric composition	Mass spectrometer
Inorganic analysis	X-ray fluorescence spectrometer
Meteorology	Pressure, temperature, and wind velocity sensors
Seismology	Three-axis seismometer
Magnetic properties	Magnets on sampler and a camera test chart, observed by cameras
Physical properties	Various engineering sensors
Radio science	
Celestial mechanics	Orbiter and lander radio and radar systems
Atmospheric properties	
Test of general relativity	

The main functions of the orbiter cameras, whose pictures are displayed in this book, were to aid in the selection of safe landing sites, to establish the geologic and dynamic environments in which the lander experiments were performed, and to add to our knowledge of the evolution of the Martian surface and the dynamics of its atmosphere.

Mariner 4, launched in 1964, weighed only 262 kg and carried only a small science payload. As it flew by Mars at a minimum range of slightly less than 10 000 km, it returned 21 pictures of the planet. An airbrush mosaic of two of the better Mariner 4 frames is shown here. The resolution is approximately one kilometer, about 150 times better than Earth-based photographs. The pictures show a cratered surface that superficially resembles the Moon, a result that was somewhat disappointing in view of some of the more extreme expectations. Certainly, no canals were observed.

In 1969, twin spacecraft, Mariner 6 and Mariner 7, were sent to examine Mars. Like Mariner 4, these spacecraft were flybys, and between them they acquired 201 images that were a blend of low-resolution, wide-area frames and nested, high-resolution frames. A typical image pair is shown in the frames taken south of Sinus Sabaeus. The highest resolution achieved was approximately 500 meters. Again, most of the area photographed resembled the Moon, and craters dominated the landscape. Only a few pictures displayed nonlunar-like features, such as chaotic and featureless terrains, that might have hinted at what remained to be discovered. In fact, the two spacecraft had passed over only those parts of the planet that retain an ancient cratered surface, and they had missed the parts of the planet that have younger and more diverse geological features.

Mariner 9, the final predecessor to Viking, eventually revealed the extraordinary diversity of the planet's surface. Arriving at Mars in November 1971, Mariner 9 was the first spacecraft to go into orbit about another planet. It took more than 7300 images of Mars, covered the entire surface at a resolution ranging from 1 to 3 km, and covered selected areas at resolutions down to 100 meters. The typical Mariner 9 image included here was taken in the same area as the Mariner 4 images. Mariner 9 operated for almost a year, nearly four times the minimum mission requirement, before running out of attitude control gas. Along with the images, extensive data were obtained on the atmosphere, surface temperature, and global weather patterns.

The images from Mariner 9 were the most exciting ever obtained in planetary exploration, revealing giant canyons and volcanoes, large channels (possibly cut by liquid water), and puzzling features that defied all geological explanations. The stage was set for the Viking missions.

Viking 1 was launched from Kennedy Space Center at Cape Canaveral on August 20, 1975, and arrived at Mars on June 19, 1976. Initially, the spacecraft was put into a Mars-synchronous elliptical orbit with a period of 24.66 hours, an apoapsis of 33 000 km, and a periapsis of 1513 km. During the first month, Viking 1 was used exclusively to search for and certify a safe landing site for Viking Lander 1. After the lander touched down on Mars on

July 20, 1976, (the seventh anniversary of the first manned lunar landing) the orbiter began systematically imaging the surface. Its highly elliptical orbit was particularly suited for studying the surface because it allowed a mix of close-up, detailed views at periapsis and long-range, synoptic views near or at apoapsis.

Table 1 is a chronology of the orbit of Viking Orbiter 1. Two events merit further description. On February 12, 1977, the orbit was changed to permit a flyby of Phobos, the larger, inner Martian moon. At closest approach, the orbiter flew within 90 km of the surface of Phobos. On March 11, 1977, the periapsis of Viking Orbiter 1 was lowered to 300 km from the Martian surface. At this low periapsis, surface features as small as 20 meters across could be identified, while the ability to acquire lower resolution and greater areal coverage away from periapsis was retained. At the beginning of this decade, Viking Orbiter 1 had taken more than 30 000 pictures of the planet and was still operational.

TABLE 1.—Viking Orbiter 1 Chronology

Date	Revolution	Event
June 19, 1976	0	Mars orbit insertion
June 21, 1976	2	Trim to planned site-certification orbit
July 9, 1976	19	Orbit trim to move westward
July 14, 1976	24	Synchronous orbit over landing site
July 20, 1976	30	VL-1 landing at 1153:06 UTC
Aug. 3, 1976	43	Minor orbit trim to maintain synchronization over VL-1
Sept. 3, 1976	75	VL-2 landing
Sept. 11, 1976	82	Decrease of orbit period to begin eastward walk
Sept. 20, 1976	92	Orbit trim to permit synchronization over VL-2
Sept. 24, 1976	96	Synchronous orbit over VL-2
Jan. 22, 1977	213	Period change to approach Phobos
Feb. 4, 1977	227	Orbit synchronization with Phobos period
Feb. 12, 1977	235	Precise correction to Phobos synchronization
March 11, 1977	263	Reduction of periapsis to 300 km
March 24, 1977	278	Adjustment of orbit period to 23.5 hours
May 15, 1977	331	Small Phobos-avoidance maneuver
July 1, 1977	379	Adjustment of orbit period to 24.0 hours
Dec. 2, 1978	898	Adjustment of orbit period to 24.85 hours; beginning slow walk around planet
May 19, 1978	1061	Adjustment of orbit period to 25.0 hours; acceleration of walk rate
July 20, 1979	1120	Raising of periapsis to 357 km; adjustment of orbit period to 24.8 hours; and slowing of walk rate

Viking 2 was launched September 9, 1975, and arrived at Mars on August 7, 1976. Like its predecessor, Viking 2 spent nearly a month after arrival finding and certifying a landing site for its lander. Table 2 is a chronology of Viking Orbiter 2. A major difference in the orbit of this spacecraft compared to that of Viking Orbiter 1 is its high inclination, which allowed Viking Orbiter 2 to observe the complex, enigmatic polar regions at relatively close range. Later in its mission, Viking Orbiter 2 flew by Deimos, the smaller of the two Martian moons, at a distance of only 22 km. Spectacular pictures showing features as small as a compact car were taken. Viking Orbiter 2 returned nearly 16 000 pictures of Mars and its satellites before it was powered down on July 25, 1978.

TABLE 2.—Viking Orbiter 2 Chronology

Date	Revolution	Event
Aug. 7, 1976	0	Mars orbit insertion
Aug. 9, 1976	2	Period and altitude adjustment; beginning of westward walk
Aug. 14, 1976	6	Increase of period to increase walk rate
Aug. 25, 1976	16	Decrease of walk rate to proceed to landing site
Aug. 27, 1976	18	Synchronous orbit over landing site
Sept. 3, 1976	25	VL-2 landing at 2237:50 UTC
Sept. 30, 1976	51	Change of orbit plane to 75° inclination and beginning of westward walk
Dec. 20, 1976	123	Lowering of periapsis to 800 km and increasing of inclination to 80°
March 2, 1977	189	Synchronous orbit over VL-2
April 18, 1977	235	Period change: 13 revolutions equals 12 Mars days
Sept. 25, 1977	404	Change of orbit period to approach Deimos
Oct. 9, 1977	418	Orbit synchronization with Deimos
Oct. 23, 1977	432	Change of orbit period to 24.0 hours and lowering of periapsis to 300 km
July 25, 1978	706	Powered down

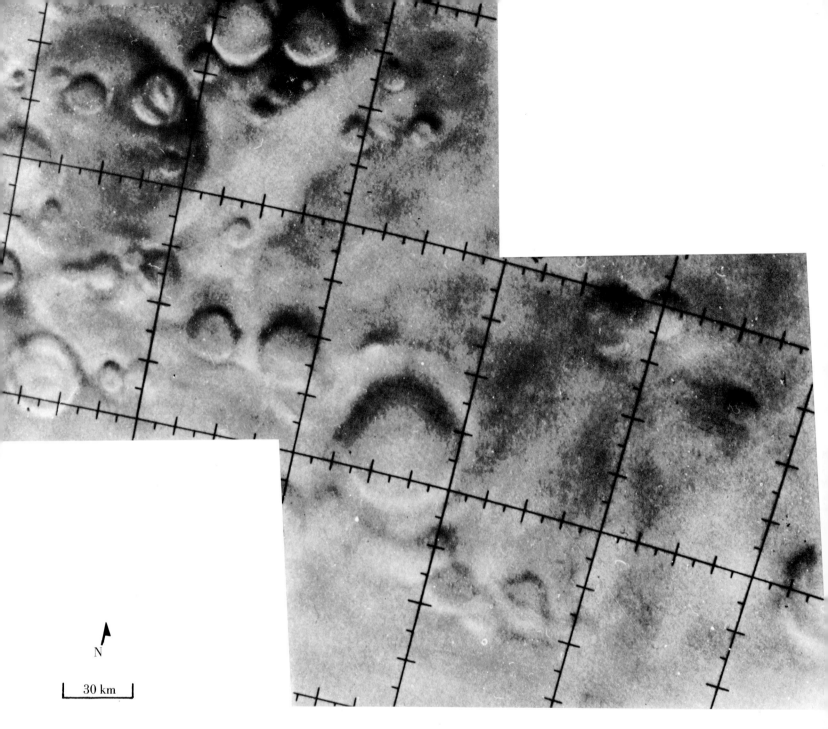

Mariner 4 Mosaic of Southern Hemisphere Ancient Cratered Terrain. This mosaic of two Mariner 4 frames suggests a lunar-like surface and shows no traces of the canals and oases of early observers nor the volcanoes, canyons, and channels revealed by later missions. The major surprises to come from Mars remained hidden. [Mariner 4 frames 15-16A; 47° S, 141° W]

Mariner 6 Frames South of Sinus Sabaeus. (a) A low resolution frame from Mariner 6 shows the large craters Wislicenus (left) and Flaguergues (right). (b) This is a high resolution frame of a small portion of the northeast rim of Flaguergues and the adjoining terrain. Fortuitous placing of all the Mariner 6 and 7 frames within the oldest terrain of Mars was such that none of the more spectacular younger features was observed. [Mariner 6 frames (a) 6N21, (b) 6N22; 15° S, 345° W]

◁

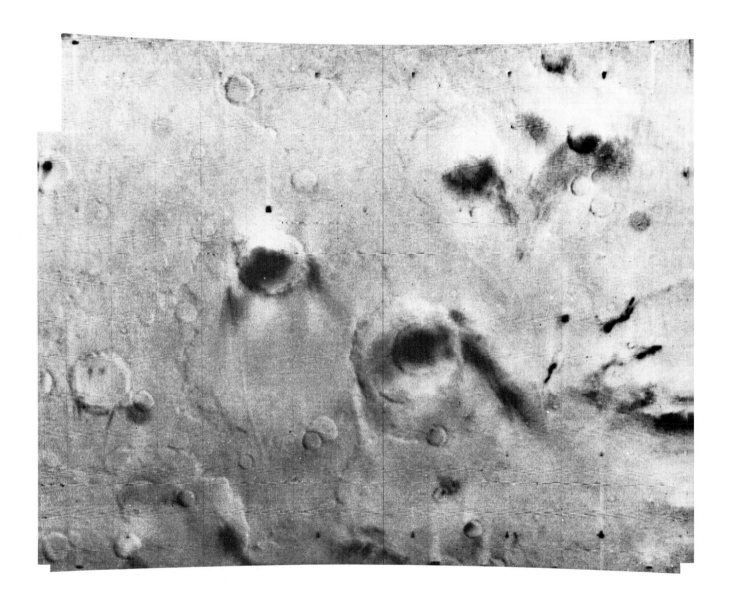

Mariner 9 Frame of Southern Hemisphere Ancient Cratered Terrain. A typical Mariner 9 image in the same area as shown for Mariner 4 confirms the lunar-like appearance of most of the southern hemisphere of Mars. Mariner 9 made many discoveries about Mars, including channels, giant volcanoes, and a giant rift valley, which set the stage for the Viking missions. [Mariner 9 frame 115A18/32, DAS time 05706928; 49° S, 141° W]

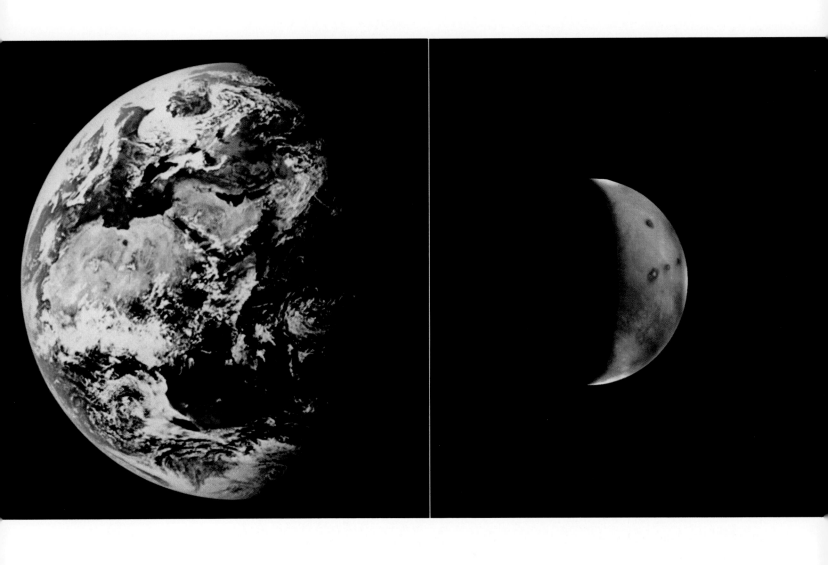

EARTH AND MARS:
A COMPARISON

MARS, the outermost of the four terrestrial planets (Mercury, Venus, Earth, and Mars), is the second closest to Earth after Venus. It is slightly more than half the size of Earth and almost twice the size of the Moon. The atmosphere is very thin, less than one one-hundredth that of Earth, and composed primarily of carbon dioxide.

Temperatures are cold, the mean annual surface temperature being approximately -50°C at the equator and close to -130°C at the poles. Because of the thin atmosphere, the diurnal temperature range is large, greater than 100°C at the equator. Summer temperatures rise above 0°C at midday despite the low diurnal mean.

The rotational axis is inclined to the ecliptic, like Earth's, so that Mars experiences distinct seasonal weather patterns. A particularly striking seasonal event is the annual dust storm. During summer in the southern hemisphere, large dust storms develop and obscure much of the planet's surface from view. Such storms, long known from telescopic observations, were observed from orbit by Mariner 9 and both Viking orbiters. During the height of the 1977 dust storm season, wind speeds up to 25 meters/sec were recorded by the Viking landers, although they were far from the center of dust storm activity.

Another regular seasonal event is the formation of clouds of carbon dioxide ice particles in the polar regions during the fall as gas starts to condense out of the atmosphere onto the growing cap. So much of the atmosphere condenses out in this process that atmospheric pressure decreases more than 30 percent from fall to winter. The pressure decrease is smaller in northern winter because of the smaller northern cap.

Other cloud activity is related to water in the atmosphere. Although very small amounts of water are present, the atmosphere is close to saturation much of the time, and a wide variety of water ice clouds have been observed.

The Martian surface has some characteristics of Earth's surface, some of the Moon's, and some unique features. The planet is very asymmetric in appearance. Most of the southern hemisphere is densely cratered and superficially resembles the lunar highlands. In contrast, the northern hemisphere is relatively sparsely cratered and has many large volcanoes that have no lunar counterparts.

The most prominent volcanic region is Tharsis, where there are several very large volcanoes that resemble terrestrial shield volcanoes, such as those in Hawaii, except that those on Mars are many times larger. The different features of the volcanoes, such as calderas, lava flows, and lava channels, are also many times larger than their terrestrial counterparts.

Tharsis is close to the center of a 6000-km diameter, 7-km-high bulge in the Martian crust. Numerous fractures radiate from the center of the bulge and extend out as far as 4000 km. The fractures are arrayed unevenly and are concentrated in intensely fractured zones called fossae. The radial fractures are so extensive that they are the dominant structural element over half the planet's surface.

Close to the equator, east of Tharsis, a series of vast interconnected canyons constitutes the Valles Marineris. These canyons also are enormous by terrestrial standards, being almost 4000 km long, up to 250 km across, and up to 9 km deep. The canyons are aligned radial to the center of the Tharsis bulge, and appear related in some way to the radial fractures.

To the east, the canyons become shallower and merge into a type of terrain peculiar to Mars: Large areas of the surface apparently collapsed to form arrays of jostled blocks that are at a lower elevation than the surrounding terrain. Because of the jumbled nature of the surface, this terrain has been termed chaotic.

From many of the regions of chaotic terrain, large, dry river beds emerge. The channels generally start full size and extend down the regional slope for several hundred kilometers. Most large channels emerge from the chaotic terrain just east of Valles Marineris and flow into the Chryse basin to the north, but several occur elsewhere. In addition to these very large channels, numerous smaller tributary systems and dendritic drainage networks are present throughout the equatorial regions.

The origin of the channels is not known, and they have been the subject of a lively debate since their discovery in 1972. The main issue is whether or not they were formed by running water. If they were, then different climatic conditions in the past may be implied. There are also some intriguing biological implications if "wet" periods have occurred in the planet's history.

The effects of wind are evident over almost all the Martian surface. Many of the classic dark markings apparently are associated with wind activity. They can commonly be resolved into arrays of streaks that start at craters and are aligned parallel to the predominant winds. The streaks may be lighter or darker than the surroundings.

Wind action is also evident from the streamlined form of many features. High resolution pictures show that the north pole is almost entirely surrounded by dune fields that form a dark collar around the pole. Dunes also occur elsewhere, such as in the canyons and within craters, especially in high southern latitudes. Thick sequences of layered deposits of unknown origin are found at both poles. They lie unconformably on the terrain and appear to be very young compared with most other features on the planet. The deposits possibly are accumulations of windblown debris mixed with condensed volatiles like water.

The geologic histories of Mars and Earth are quite different, partly because of the internal dynamics of the planets and partly because of the differing effects of the atmospheres and oceans. Earth's geology is dominated by the effects of plate tectonics. The rigid outer shell of the

Earth (the lithosphere) is divided into plates that move laterally with respect to one another. Where plates diverge, as at midoceanic ridges, new crust forms; where they converge, one plate generally rides under the other to form a subduction zone. Thick sediments may accumulate in a subduction zone; these may ultimately be compressed, partially melted, and uplifted to form linear mountain chains of folded and partly metamorphosed rocks, such as the Andes and the Himalayas. Melting of the subducted plate as it moves down into the mantle may also given rise to volcanism in the subduction zone. Where plates move laterally with respect to one another, they form transcurrent faults such as the San Andreas. The present configuration of the Earth's surface is thus a partial record of the motion of different plates with respect to one another.

Mars displays little, if any, evidence of plate motion. The crust appears very stable. Long linear mountain chains and subduction zones are absent, and transcurrent faults and compressional features of any kind are rare. Its geologic history is thus very different from that of the Earth.

Greater stability on Mars results in the preservation of much older features. On Earth, surface materials are recycled at a relatively rapid rate by erosional processes and subduction. The two processes are commonly interdependent; for example, erosion is greatly increased in mountainous regions along subduction zones. On Mars, however, recycling of crustal materials is extremely slow, as evidenced by the preservation of large areas of old, densely cratered terrain that probably dates back approximately four billion years.

Crustal stability may also be the cause of the large size of the Martian volcanoes. On Earth, volcanoes are limited in size because plate motion usually carries them away from the magma source. On Mars, however, a volcano remains over its source and can continue to grow as long as magma is available.

The preservation of features billions of years old on the Martian surface indicates extremely low erosion rates. On Earth, most erosion results from running water. Small channels in the old cratered terrain of Mars are evidence of an early period of fluvial action, but survival of the old craters indicates that the period was short. For most of the planet's history, wind has probably been the main erosive agent. Despite giant dust storms, however, the wind clearly has not been very efficient in eroding the surface, because so much old terrain survives. Most of the wind's action probably involves reworking previously eroded debris.

COMPARISON OF EARTH TO MARS

Earth		Mars
12 756 km	Diameter	6787 km
5.98×10^{24} kg	Mass	0.646×10^{24} kg
9.75 m/s^2	Gravitational acceleration	3.71 m/s^2
149.5×10^6 km (average)	Distance from Sun	227.8×10^6 km (average)
839 cal/cm^2/day	Sunlight intensity	371 cal/cm^2/sol
23° 27"	Inclination	23° 59'
24^{h}00^m	Length of day	24^{h}40^m (=1 sol)
365 days	Length of year	686 days (668 sols)
60 000γ	Magnetic field	50-100γ
1013 mb (average)	Atmospheric pressure	7 mb (average)
1	Known satellites	2

THE GREAT EQUATORIAL CANYONS

VALLES MARINERIS is composed of steep-walled canyons, individually measuring up to 9 km deep, 250 km wide, and 1000 long. They were named for Mariner 9, the Mars-orbiting spacecraft that took the first pictures of the canyons in 1971. The entire Valles Marineris system extends over 4000 km from west to east near the Martian equator, and its size dwarfs all similar terrestrial features except, perhaps, the 5000-km-long midocean rift system.

Pictures taken by the Viking orbiters show large areas of Valles Marineris at a resolution far better than that achieved by Mariner 9. The new features observed indicate that, although erosional landforms (such as landslide scars and deposits) and tributary canyons are common, faulting apparently has been the dominant factor in canyon development.

Since discovery of Valles Marineris, the method of their formation has been a nagging puzzle. The canyons do not form a well-integrated drainage system; some are completely closed depressions, and lateral transport by wind or water would be considerably impeded. Now, however, the new evidence of faulting suggests that most negative relief results from subsidence. Low, straight scarps, which apparently indicate downward subsidence of canyon floors along faults, cut across erosional features on many canyon walls. Similar scale faulting occurs on Earth: in East Africa the continental crust is in tension across large rift valleys. Erosion of the Valles Marineris walls apparently continued into the recent past, so the crustal tension causing the faulting within the canyons may also have been a relatively recent phenomenon.

Another exciting discovery resulting from Viking images is the presence of thick layered deposits on the floors of several canyons. Layered rock is also visible in the canyon walls, and thus is part of the Martian crust predating the canyons. Some materials on the canyon floors are distinctive for the fine scale and regularity of their layering. Only climatic modulation of a sedimentary process seems adequate to explain them. Possibly, regular changes in the Martian climate, governed by known orbital variations, have controlled the level of dust storm activity and the rate of deposition of sediment from the atmosphere. Another theory is that some sections of Valles Marineris were sites of lakes in which layered sediments were deposited. Before Viking, regularly layered deposits were known only in the polar regions of Mars, and their creation may also be associated with cyclical climatic change.

Impact craters, which are so numerous on other Martian terrains, are scarce within Valles Marineris. They appear most frequently on smooth areas of canyon floor, and are possibly the tops of blocks down-faulted from the upland plain. Shallow pits, possibly eroded impact craters, are abundant in other places. Impact craters are probably scarce in the canyons because erosion and deposition by landslides and wind have been actively renewing interior surfaces. No evidence of flow of water has been found within Valles Marineris, although some channels on the adjacent upland are abruptly truncated by steep canyon walls.

Global View of the Valles Marineris. As Viking 1 approached Mars in June, 1976, it recorded this color picture showing Valles Marineris stretching more than 4000 km across the face of Mars. North is toward the upper right, and it is the winter season in the southern hemisphere. The annual southern ice cap extends up to 45° S latitude, blanketing the Argyre basin, an impact crater 800 km across. [IPL, ID I 2038MBVI, VERS2; 30° S, 70° W]

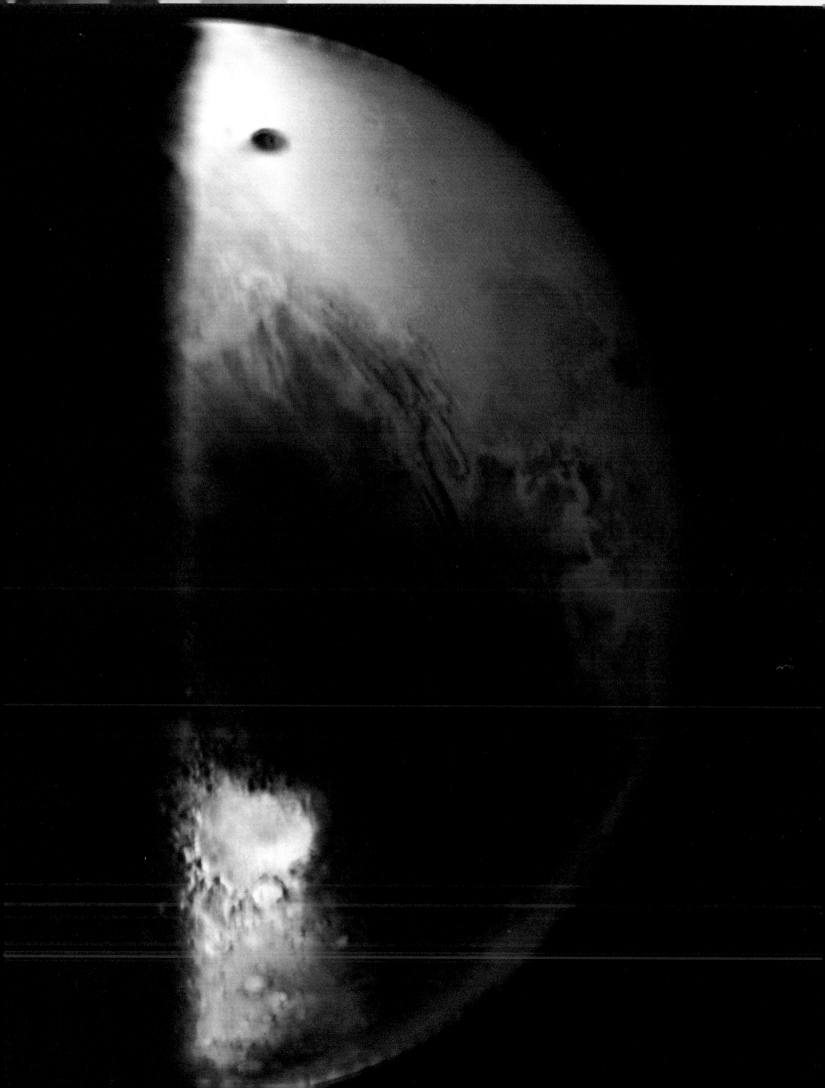

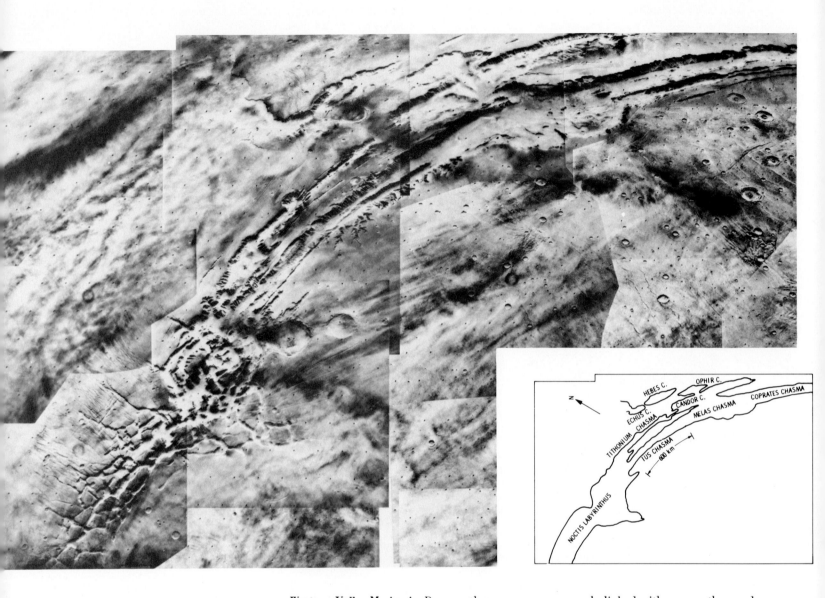

Western Valles Marineris. Because these canyons are poorly linked with one another, and their floors not a regularly graded slope, they could not have formed as water drainage features. The straight alignments of many canyon walls, and the faulting in several directions associated with Noctis Labyrinthus, combine to suggest that the Valles Marineris system is composed of great rift valleys formed on the surface of a dome. At the summit of the dome, near the labyrinth, the crust was stretched in all directions to form a network of fault-bounded valleys. On the flanks of the dome, the greatest strains were concentric about the summit, giving rise to a set of radial rift valleys. The inset shows names of individual canyons (chasma). [40A37-52; 13° S, 87° W]

Eastern Valles Marineris. The broadest valleys merge with patches of chaotic terrain, apparently the product of collapse and erosion of ancient cratered terrain. Channels beginning at the margins of chaotic terrain extend northeastward onto Chryse Planitia, the region in which Viking Lander 1 is located. The relationship between Valles Marineris and the chaos is not well understood. Irregular collapse, to form the chaos, may reflect crustal stresses similar to those forming the rift valleys, but differing in orientation and complexity. It is also possible that the chaos formed during the catastrophic release of liquid water derived from artesian reservoirs or the melting of ground ice. [32A11-15; 9° S, 53° W]

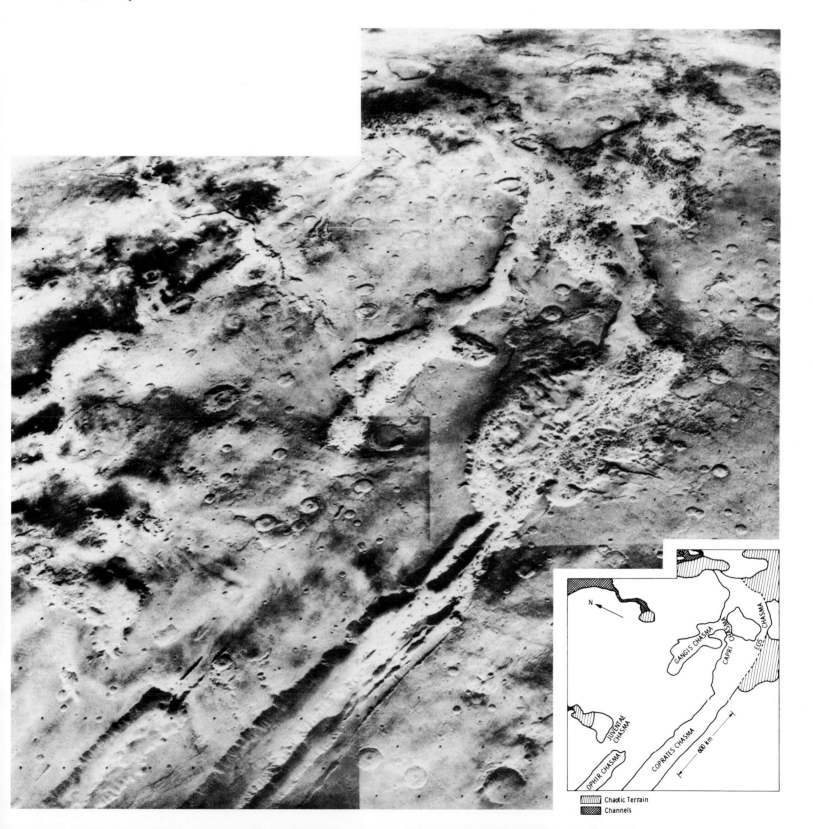

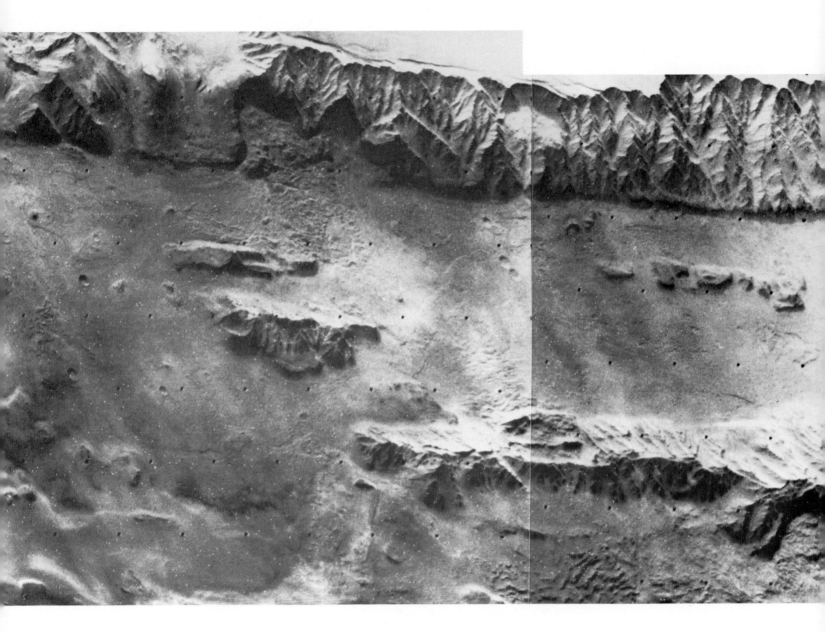

N

| 20 km |

View of Coprates Chasma Looking East-Southeast. Interplay of faulting and erosion within Valles Marineris is apparent here. A low, steep, and generally straight scarp occurs at the foot of the north canyon wall. The scarp apparently results from downward offset of the flat canyon floor relative to the wall. Gullies on the canyon wall are truncated by the scarp and thus predate it. In contrast, large-scale landsliding from two great canyon wall alcoves postdates the latest down-faulting; landslide debris has buried the scarp for a distance of about 40 km. [58A89-91; 15° S, 60° W]

Evolution of Canyons by Landsliding. This mosaic of Viking Orbiter 1 pictures was taken looking southward toward the junction of Gangis and Capri Chasmae. Rugged terrain consisting of numerous tilted blocks occurs within alcoves in the canyon wall. Extending from the alcoves are thin blankets of material with fan-like patterns of surface striations. Where two patterns intersect, one thin lobe clearly overlies the other and cuts off its striations. These features are giant landslide deposits that formed when sections of canyon rim collapsed. The broad, thin lobes of material apparently flowed at high velocity from the bases of the collapsing masses. The mechanism by which the surface striations formed is not well understood, although similar features have been observed on terrestrial landslides. Groups of hills, similar to chaotic terrain, and sand dunes contribute to a varied canyon floor landscape. Layering in the upland rocks is evident near the top of the south canyon wall within the landslide alcoves. [14A29-32; 9° S, 42° W]

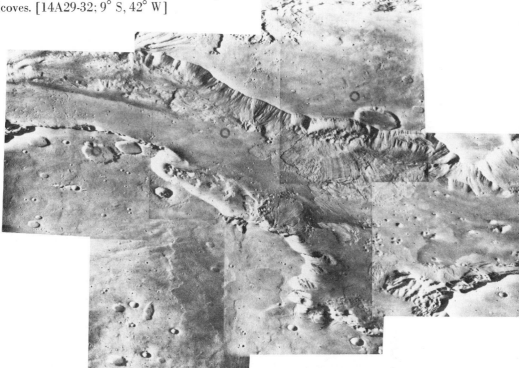

Stereogram of Ius Chasma. The south canyon wall here is incised by several large, branching tributary canyons that appear to have developed headward along joints or zones of weakness in upland rocks. It has been suggested that these canyons developed through a sapping process, in which the melting of ground ice caused the upland to collapse. The floor of the main canyon between A and B appears to have been fractured and down-faulted below the level of the mouths of the tributary canyons. Layering can be seen in several places on the canyon walls (C and D). [Left 66A07-09, Right 63A40; 7° S, 85° W]

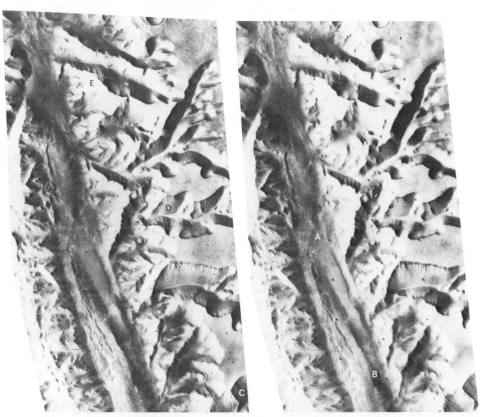

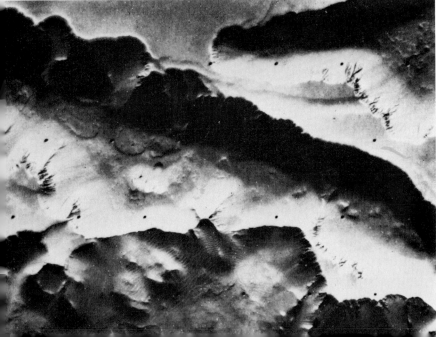

Noctis Labyrinthus. The origin of the canyons by faulting is most apparent in Noctis Labyrinthus at the western end of Valles Marineris. Many canyons have a classic graben form, with the upland plain surface preserved on the valley floor. Other canyons are more irregular in form and have rough floor terrains, evidently the consequence of landsliding and the puzzling process of pit formation. In places it appears that surface materials have sifted downward into a gaping hole in the subsurface. The inset shows a slope covered with light albedo dunes and several small landslide lobes. [46A13-28, 47A17-28, 48A21-28, 49A22-28, 50A14-28, Inset 62A64; 7° S, 100° W]

◁

Stereogram of Central Tithonium Chasma. This section of Tithonium Chasma is about 6 km deep. Overlapping landslide lobes cover the canyon floor and scarps that bound a rift valley within the canyon. On the south canyon wall, distinct bright and dark horizontal stripes are probably outcrops of layered rocks. Parallel chains of pits and graben mark the upland surface to the south. [Left 57A45, Right 64A22; 5° S, 85° W]

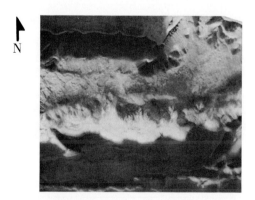

Large Albedo Contrasts and Relief in Tithonium Chasma. Large contrasts in the brightness of surface materials can so confuse perception of depth in single frames that stereoscopic imaging is necessary to interpret surface features. To the left is a long, narrow rift valley about 5 km deep. North of this valley the canyon floor is 1 to 2 km higher and irregularly mottled. A mountain with finely gullied flanks rises about 2 km from the canyon floor. This mountain is representative of many plateaus, ridges, and hills on the floors of broader canyons. They differ drastically from canyon wall materials in their patterns of erosion. Many are composed of materials with a distinct horizontal layering. [Left 44A27, Right 63A63; 5° S, 65° W]

West Candor Chasma. Here the canyon floor is entirely covered with eroded, layered materials. Layering is most prominent at A, B, and C. An offset spur (D) and a low, steep scarp (EF) along the western walls of the canyon may have been formed by faults trending north to south across the main Valles Marineris. The tributary canyon at G seems to have developed by two processes; subsidence of a block of crust (graben), and irregular collapse into a string of pits. [65A25, 57; 66A17-27; 7° S, 75° W]

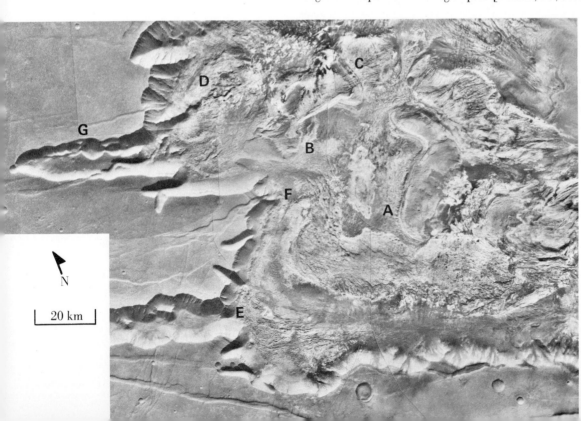

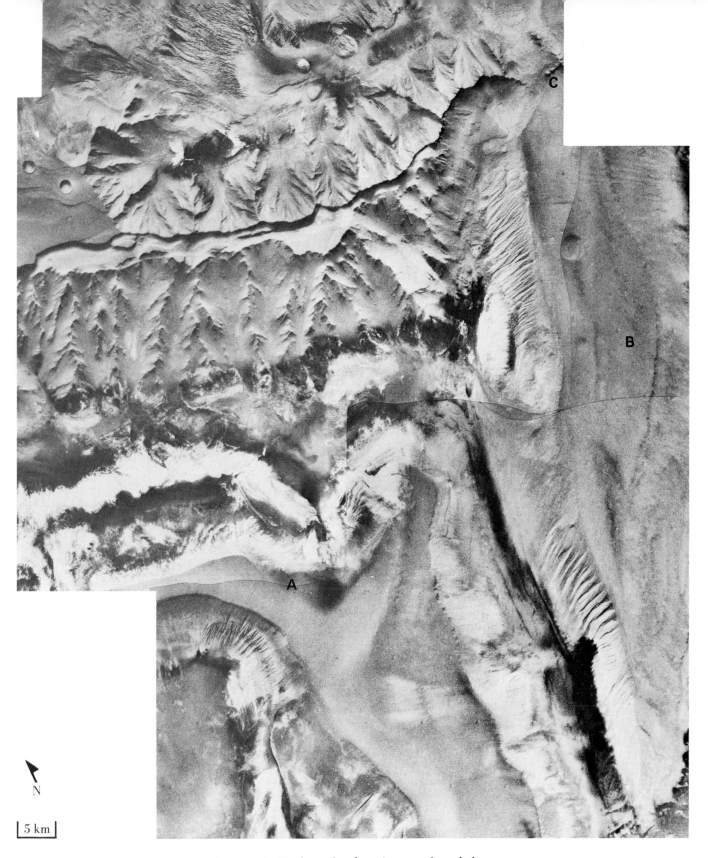

Candor and Ophir Chasmae. Large plateaus (A, B), formed at least in part of regularly layered materials (as at A), rise from the floor of Candor and extend across the gap between Candor and Ophir Chasmae. At C, plateau materials apparently were deposited upon an eroded spur of the canyon wall and are now themselves being eroded away. Streamlined ridges and grooves in Ophir Chasma are probably wind sculpted. [66A23-30; 5° S, 73° W]

Plateaus between Ophir Chasma and Candor Chasma. These enormous, streamlined plateaus bridge the gap between the two canyons, possibly indicating wind erosion on a very large scale. Alternately, it has been suggested that a lake (or sea) once existed in the north section of Ophir Chasma until the canyon wall was breached southward to unleash an enormous flood. The dark material on the canyon floors is probably a wind deposit, as dune-like forms are visible in other images. [IPL ID, IV2515CGX2, I2515DGX2; 5° S, 73° W]

◁

Layered Material in Juventae Chasma. A ridge of very uniformly layered light and dark materials rises from the floor of Juventae Chasma. Cyclical changes in sedimentation, perhaps modulated by climate, seem the most probable explanation for their origin. [81A15-17; 5° S, 62° W]

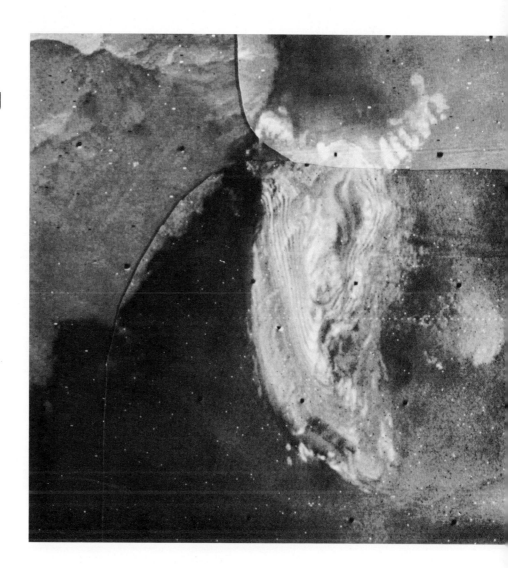

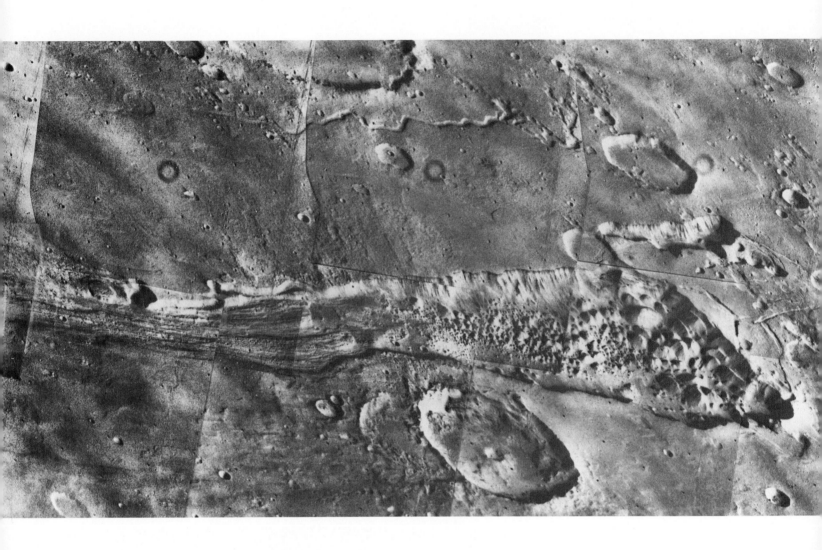

CHANNELS

CHANNELS are among the more puzzling and intriguing features of the Martian surface. The most controversial aspect of the channels is whether they were formed by running water. Present climatic conditions on Mars prevent the existence of liquid water at the surface, so a water-worn origin implies that very different climatic conditions prevailed in the past. A denser atmosphere and higher temperature are both required. Because of the difficulty in explaining how climatic conditions could have changed so drastically, alternative methods of erosion, such as by wind and lava, have been suggested.

Three main types of channels have been recognized: (1) runoff channels appear as dendritic networks, or arrays of relatively small channels or valleys located mainly in the old, densely cratered terrain; (2) outflow channels appear as large scale tributaries; and (3) fretted channels appear as long, relatively wide, flat-floored valleys that possess tributaries and increase in size downstream.

Much of the old, cratered terrain, particularly in the equatorial regions, is dissected by channels of some type, the most common of which is the simple gully, typically a few tens of kilometers long. The gullies generally have few tributaries which, if present, have small junction angles. The numerous well-integrated tributary networks provide the strongest evidence for water erosion at some period in Martian history, because they are unlikely to form by wind action or lava erosion. The runoff channels are largely restricted to the oldest terrain and are themselves commonly degraded. Most of these channels therefore appear to have formed early in the planet's history.

Most outflow channels occur around the Chryse basin. They commonly emerge full size from chaotic terrain that has seemingly collapsed to form areas of jostled blocks as many as 3 km below the surroundings. The channels extend from the chaotic terrain downslope several hundred kilometers into the plain of Chryse Planitia. They may be tens of kilometers wide and more than a kilometer deep, a size indicating erosion on an enormous scale.

Within the channels are many features, such as teardrop-shaped islands, longitudinal grooves, terraced margins, and inner channel cataracts, that are also found in regions on Earth affected by large floods. The dimensions of the Martian channels suggest peak flood discharges of 10^7 to 10^9 m^3/sec. By comparison, the average discharge of the Amazon is 10^5 m^3/sec, and the largest known terrestrial flood, the Lake Missoula flood that occurred in eastern Washington in the late Pleistocene, had a peak discharge of

10^7 m^3/sec. Thus the Chryse outflow channels, and similar ones elsewhere, provide evidence of enormous floods on Mars—far greater than any known on Earth.

The time period in which the climatic conditions permitted liquid water to exist is uncertain because of the difficulty of precisely dating the channels. Most of the evidence, however, suggests that the more clement conditions prevailed very early, perhaps during Mars' first billion years, and that this period was followed by general global cooling. The present harsh conditions have probably existed for most of the planet's history.

Fretted channels occur mostly within the old, densely cratered terrain, especially at its boundaries with younger units. They lack features indicative of catastrophic flooding. The presence of tributaries and a decrease in channel size upstream also argue against formation by floods. The origin of the fretted channels is not known, but numerous features on the floors suggest that mass wasting may have played a significant role.

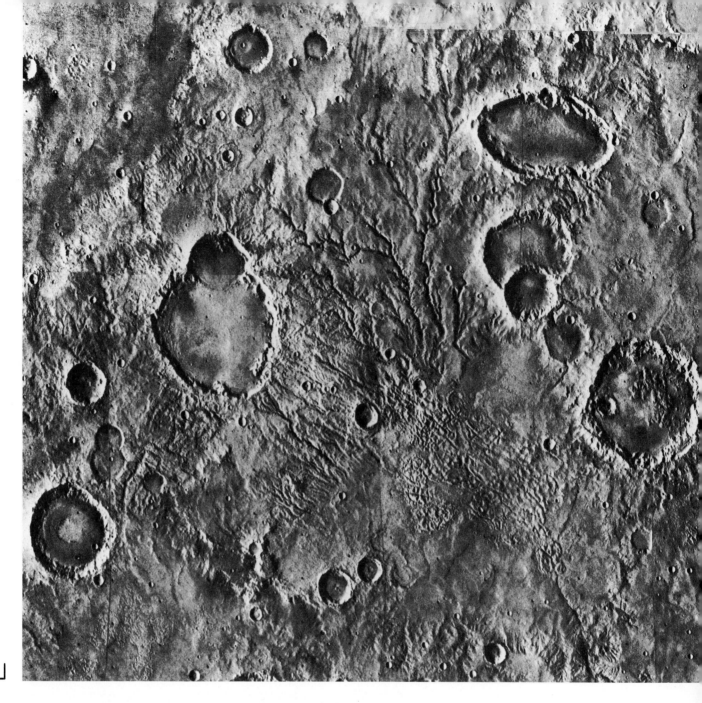

Dendritic Channels in the Southern Highlands. These channels are deeply incised with relatively wide, undissected interfluves. Most terminate abruptly at their lower ends. [P-18115; 25° S, 10° W]

◁

Finely Channeled Old Cratered Terrain. The channels are concentrated on crater rims and tend to be approximately parallel, a few tens of kilometers long, with few tributaries. Such channels are typical of much of the heavily cratered terrain of Mars, but are rare in the sparsely cratered areas. [84A16-22; 23° S, 0° W]

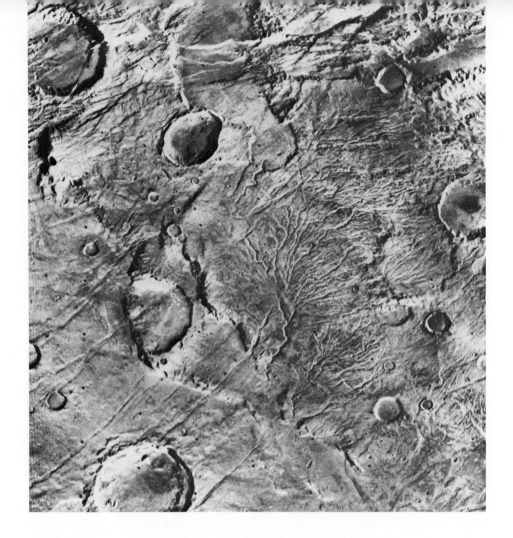

Dense Drainage Network in the Southern Highlands. Dendritic patterns like these suggest a fluvial origin and argue against alternatives, such as erosion by wind or lava. [63A09; 48° S, 98° W]

Outflow Channel Emerging from Chaotic Terrain. Oblique view, looking south, of the source region of an apparent flood. The channel starts full scale in a region of chaos enclosed by cliffs. Possible mechanisms for producing such a relation are rapid release of water from buried aquifers or the melting of ground ice by volcanism. [P-16983; 1° S, 43° W]

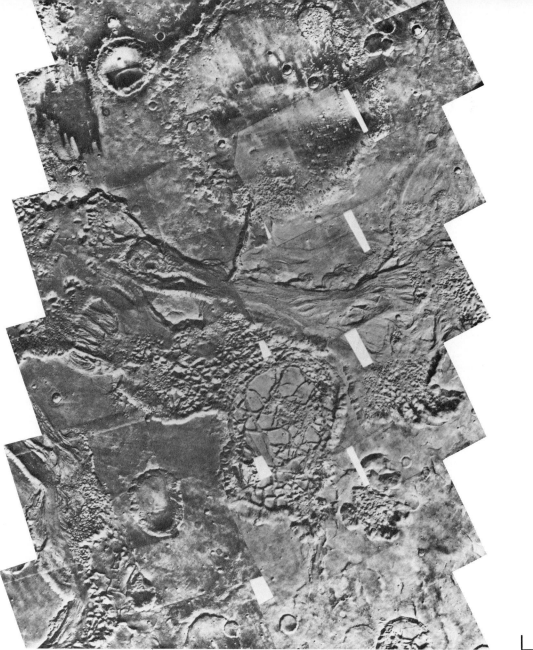

Channels and Chaotic Terrain at the Source of Tiu Vallis. A 50-km-wide channel emerges from chaotic terrain. It extends off the picture to the north, down the regional slope to Chryse Planitia 1000 km away. [P19131; 5° S, 29° W]

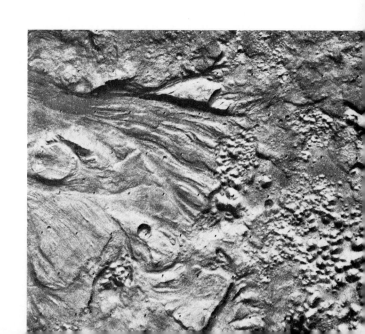

Closeup of Part of Preceding Image. This frame shows Tiu Vallis (left) extending northwest from Hydaspis Chaos (right). Hydaspis Chaos is an elongated area of collapsed terrain almost 100 km wide, from which emerge the lineated bedforms and teardrop-shaped islands of Tiu Vallis. [83A37; 3° S, 27° W]

Part of Ares Vallis. (a) This mosaic shows part of Ares Vallis incised in the heavily cratered upland. The channel is 25 km wide and about 1 km deep. Several layers, probably lava flows, are exposed in the walls of the channel. Many craters are present in the upland surface, but craters are few on the channel floor. (b) A stereogram shows the origin of Ares Vallis in chaotic terrain. Channels from two large areas of chaos have merged into a single channel. Where flow from the lower chaos has merged into the channel, many streamlined forms are visible. [(a) 211-5238; 10° N, 24° W, (b) 451A03-10; 2° N, 19° W]

 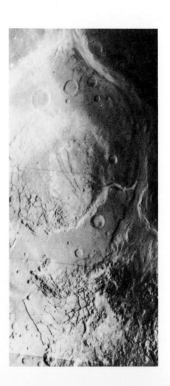

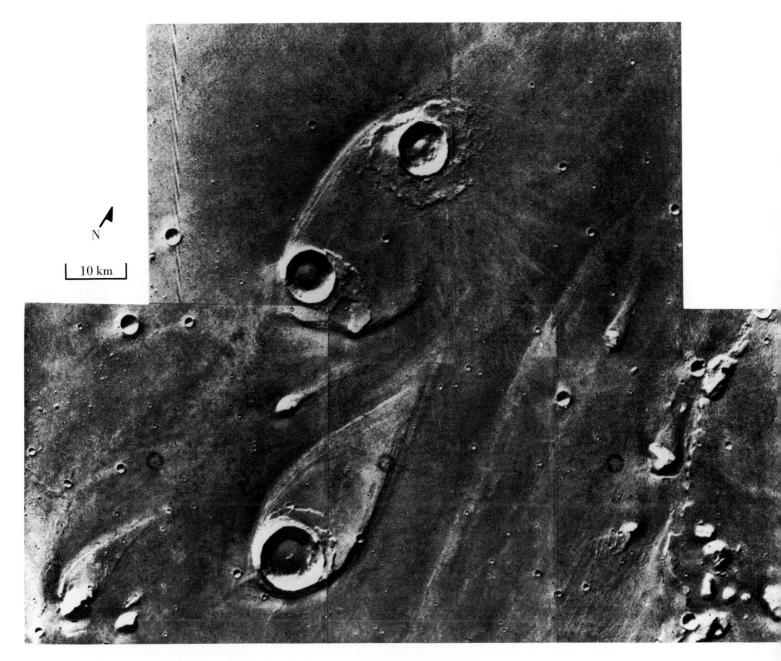

"Islands" near Chryse Planitia. Teardrop-shaped "islands" are shown at the mouth of Ares Vallis near the southern boundary of Chryse Planitia. Flow was from the south and apparently diverged around obstacles such as craters and low hills to form a sharp prow upstream and an elongate tail downstream. A shallow moat surrounds the entire island. Similar patterns on Earth have been formed by catastrophic floods, wind erosion, and glacial action. From top to bottom, the three large craters are named Lod, Bok, and Gold. [211-4987; 21° N, 31° W]

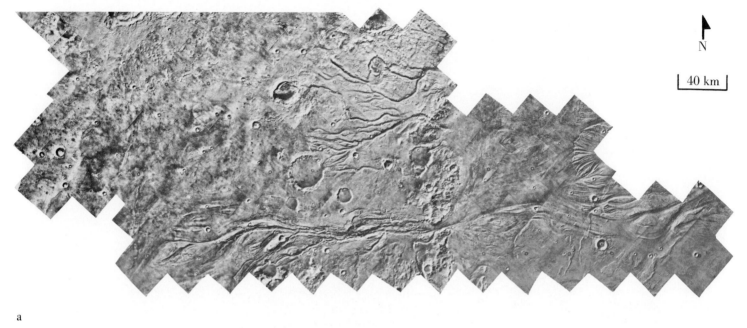

a

b

38

Channels between Lunae Planum and Chryse Planitia. (a) Channels have been cut across old cratered terrain between the lava plains of Lunae Planum on the left and the plains of Chryse Planitia to the right. Three separate channel systems are visible, starting from the north: Vedra Vallis, Maumee Vallis, and Maja Vallis. Flow along the eastern edge of Lunae Planum converged to cut Maja Vallis. Numerous teardrop-shaped islands occur upstream of the main channel. Below the channel to the east (off the right side), the flow diverges across Chryse Planitia. (b) This stereogram shows Vedra and Maumee Valles between Lunae Planum and Chryse Planitia. Note that a branch of Vedra Vallis passes through Banh Crater. [(a) 211-5190, (b) 211-5419; 18° N, 55° W]

◁

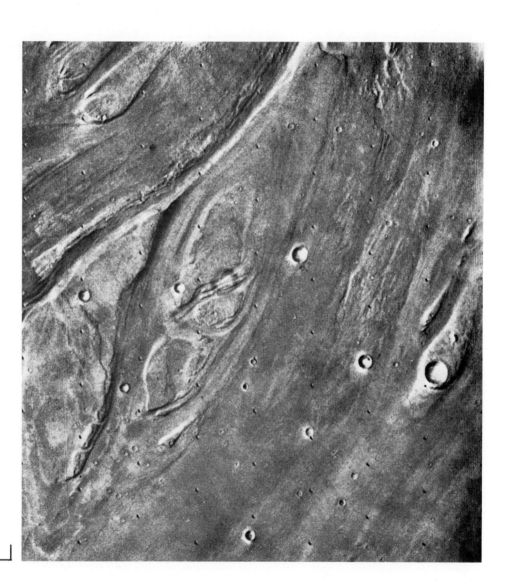

Upper Reaches of Maja Vallis. The surface of Lunae Planum is extremely scoured, with long linear grooves and teardrop islands. Flow apparently converged on Maja Vallis from a wide area of Lunae Planum. [44A44; 17° N, 57° W]

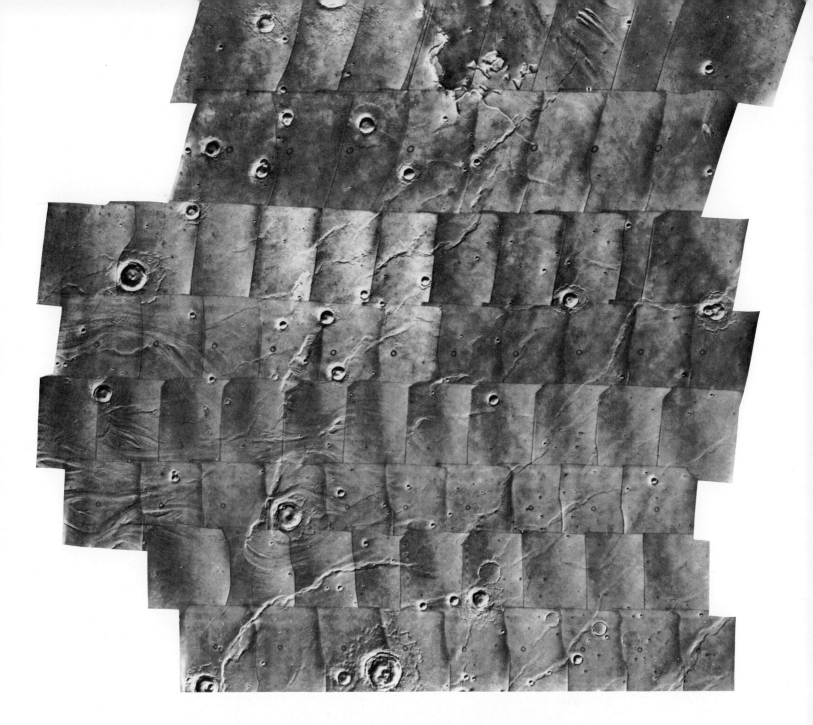

N

| 40 km |

Western Chryse Planitia. The west side of Chryse Planitia has been extensively sculpted by flow from Maja Vallis, which is situated just to the left of this mosiac. Flow diverged across the gently sloping plain of Chryse Planitia to form the sculpted features seen in this mosiac. Ridges, similar to those on the lunar maria, appear to have partly dammed or diverted flow to form a variety of scour patterns. [211-5015; 21° N, 49° W]

Flow around Dromore Crater in Chryse Planitia. Flow was from the left and apparently ponded west of the mare ridge. It then cut gaps as it flowed over low points in the ridge. Similar relations occur in the channeled scablands of Washington state. After crossing the ridge, the flow cut grooves in the Chryse Planitia floor as it flowed around Dromore, an older impact crater. [20A62; 20° N, 49° W]

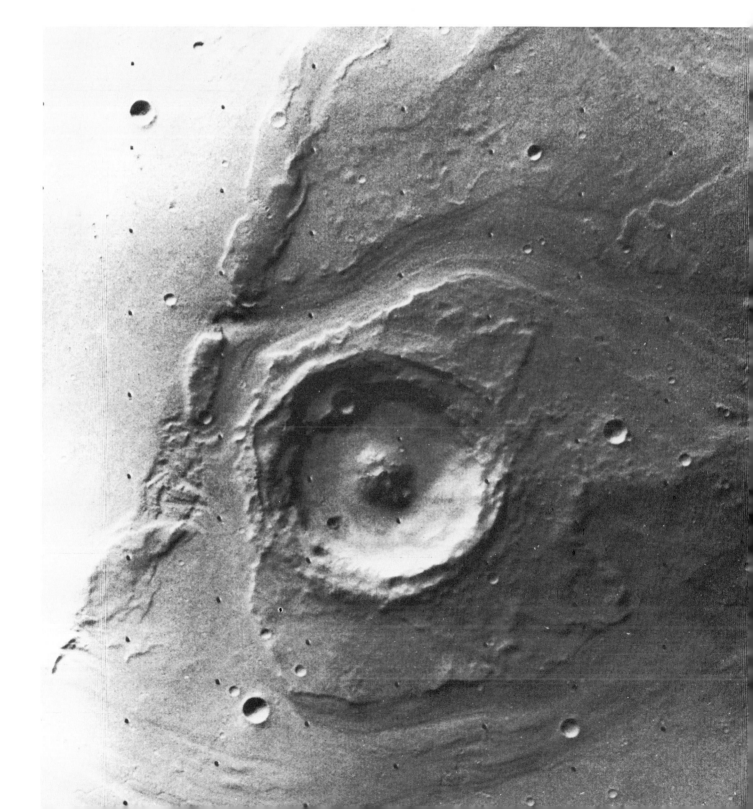

Bahram Vallis in Northern Lunae Planum. This channel has few tributaries, a flat floor, and cuspate walls, and resembles lunar sinuous rilles that are believed to be formed by lava flow. [211-5189; 20° N, 57° W]

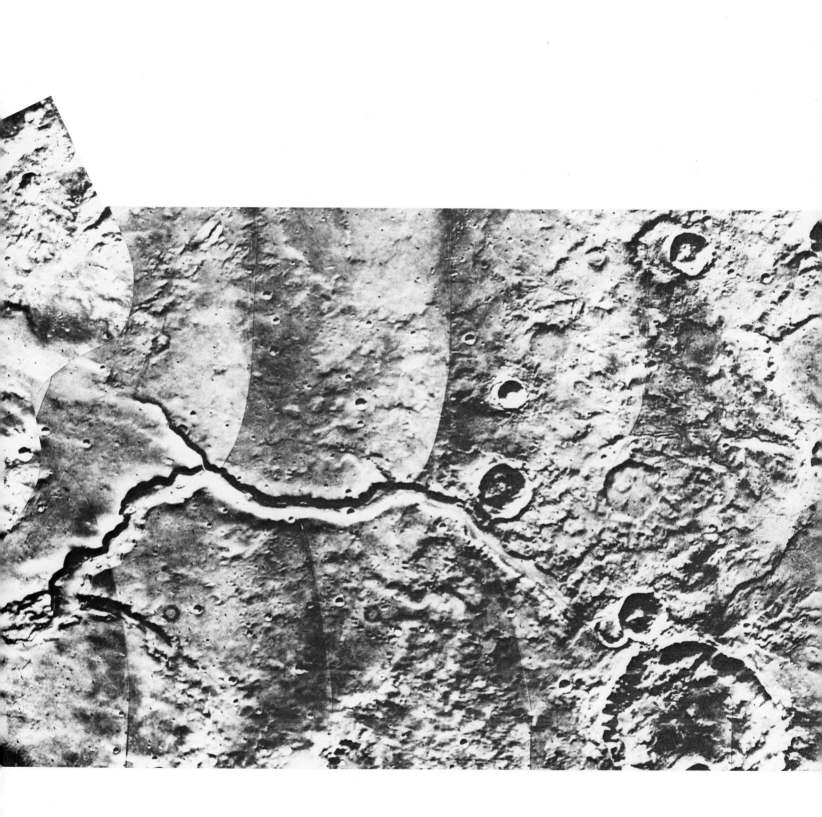

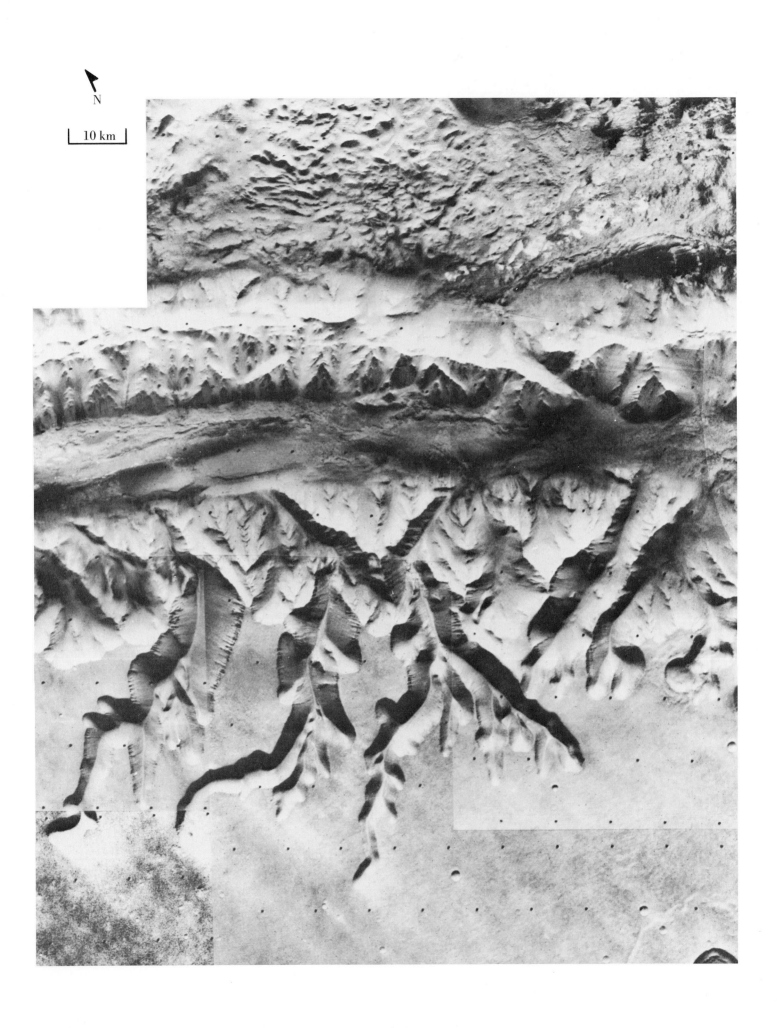

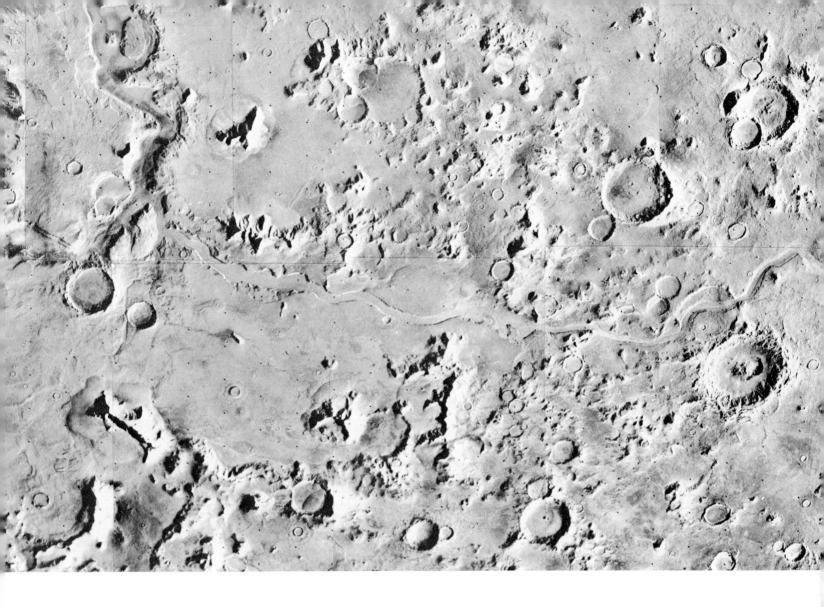

Flat-Floored Valley Northeast of Hellas. This valley is several kilometers wide and is cut into layered deposits that are clearly exposed in the valley walls. In some places, a channel is visible in the valley floor. Extensive debris fans surround many hills in the area and are probably formed by creeping of near-surface materials, perhaps aided by interstitial ice. [97A60-68; 43° S, 253° W]

◁

Section of Valles Marineris. Each tributary on the southern wall of the canyon heads in a cirque-like feature and lacks a fine-scale drainage network. The morphology suggests formation by ground water sapping rather than by surface run-off. Ground ice is a possible source for the water. [211-5158; 80° S, 85° W]

VOLCANIC FEATURES

VOLCANIC ACTIVITY on Earth can be divided into two basic types: eruptions that occur repeatedly from the same conduit and slowly build roughly circular mountains, and eruptions from any widely spaced vents, usually fissures, that create extensive lava plains. Both types are found on Mars. Volcanic rocks are of particular interest to the geologist because they originate deep within the planet and provide a means of assessing the conditions and processes that operate there. Although we are unable to examine the rocks on Mars directly, the volcanic features give an indication of rock composition. For example, silica-rich lavas tend to have higher viscosities and yield strengths than silica-poor lavas and so form differently shaped flows; volatile-rich, viscous lavas tend to produce abundant ash during eruptions, so ash deposits rather than lava flows are the predominant landform. The volcanoes are also interesting in that their shapes and sizes provide information on thermal conditions in the interior of the planet. The volcano height gives a means of estimating the depth of melting, and the degree of sagging of the crust under the weight of the volcano permits the viscosities of the crustal materials and hence the temperature profile to be calculated.

Martian volcanoes are most common in the region of Tharsis, where three large volcanoes (Ascreus Mons, Pavonis Mons, and Arsia Mons) form a northeast-southwest line. Another large volcano, Olympus Mons, is located about 1500 km northwest of the line. All four are enormous by terrestrial standards. Olympus Mons is more than 600 km across and towers approximately 27 km above the mean surface level. Alba Patera, just to the north of Tharsis, although only a few kilometers high, is 1700 km in diameter. The Hawaiian volcanoes, which are among the largest on Earth, are generally less than 120 km in diameter and 9 km above the ocean floor. Surrounding the massive Martian volcanoes are extensive lava plains and many smaller volcanoes such as Biblis Patera and Tharsis Tholus. Volcanoes occur in regions of the planet other than Tharsis, but tend to be smaller and older.

Each of the three Tharsis shield volcanoes has a caldera complex at its summit, apparently formed by repeated collapses following eruptions. On the flank of each edifice is a faint radial texture formed by numerous long, thin flows, some with central channels. The general morphology of the flows is similar to those on the flanks of the Hawaiian shield volcanoes and suggests fluid flow. Various concentric features such as terraces, breaks in slope, and lines of rimless depressions are superposed on the radial texture. On the northeast and southwest sides of each volcano, numerous pits in the shield coalesce to form alcoves that evidently were sources of enormous volumes of lava. Flows spread from these alcoves over the adjacent plains, covering the lower flanks of the volcanoes and extending several hundred kilometers from the source. Thus, eruptions from the Tharsis volcanoes formed both the volcanic edifices and the surrounding plains.

The main edifice of Olympus Mons resembles the Tharsis shields except that it is surrounded by a cliff that, at some points, reaches 6 km in height. In several places, lava has flowed over the cliff and across the surrounding plains, extending the volcanic edifice beyond the scarp. All around Olympus Mons, blocks of strongly ridged terrain extend as far as 1000 km from the scarp and constitute the so-called aureole. The origin of the aureole is unclear, but suggestions are that it is the remnant of a pre-Olympus volcano, that it consists of eroded ash-flow tuffs, or vast thrust sheets.

Alba Patera, just to the north of Tharsis, differs from the volcanoes already described. Although it is more than 1700 km across, it is only about 2 km high. Many flow features are visible on its flanks. These features are often as many as 10 times larger than their terrestrial counterparts, but otherwise show great similarity. The nature of Alba Patera's flow features again suggests fluid lavas.

Relatively featureless plains cover much of the planet's surface. The origins of most of the plains are not known. Although some may be largely aeolian and fluvial, evidence indicates that most are volcanic. The plains around the larger volcanoes have numerous flow features and are almost certainly volcanic. Other plains have ridges and rille-like features that resemble those on the Moon and so are suspected of being volcanic like the lunar maria. Where visible in section, the plains are layered, perhaps indicating interbedded materials of different origins.

▷

Olympus Mons. (a) This volcano, the largest on Mars, measures over 600 km across at the base, and is about 27 km high. It is surrounded by a well-defined scarp that is up to 6 km high. Flows drape over the scarp and extend onto the surrounding plains. In many places the scarp is associated with small block faults, indicating that faulting may have played a part in its development. Parts of the plains surrounding the volcano are characterized by ridged and grooved terrain that is faulted in places. The origin of this terrain is not known. (b) The stereogram permits a greater appreciation of the structure of Olympus Mons, especially the caldera and the scarp. [(a) 211-5360, (b) Left 211-5345, Right 211-5360; 18° N, 133° W]

a

N

50 km

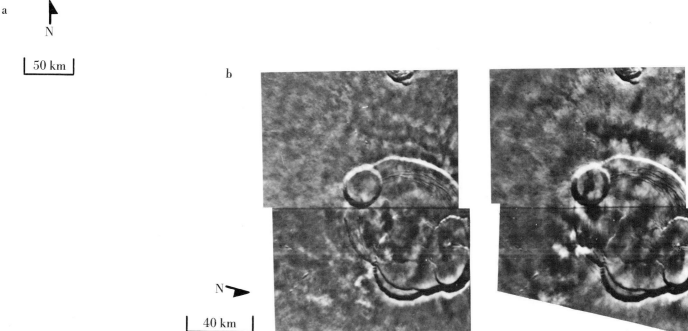

b

N

40 km

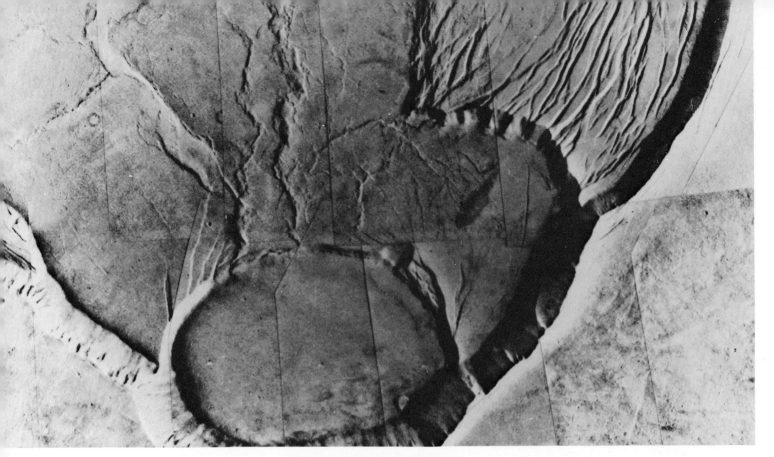

N

|—— 5 km ——|

Summit Caldera of Olympus Mons. This mosaic consists of several frames that show features on the surface as small as 18 meters across. The circular caldera on the left is almost 3 km deep and 25 km across, and has wall slopes of about $32°$. It probably formed as a result of recurrent collapse following drainage of magma out of the central conduit of the volcano during flank eruptions. The floor of the deepest caldera is featureless at this resolution, but the floor materials of other parts of the caldera complex are marked by fault patterns and ridges similar to mare ridges on the Moon. Fluting of the caldera walls suggests landslide activity. [211-5601; 18° N, 133° W]

Terraces on Upper Slopes of Olympus Mons. The origin of the lava terraces is not known. In some respects, they are analogous to terraced features seen on pahoehoe flows on Mount Etna, Sicily, where they formed as a result of embankments developing at the fronts of lava flows and the accumulation of lava lakes behind the embankments. Some of the small craters appear to be rimless volcanic pits. [46B12; 17° N, 132° W]

N

|—— 20 km ——|

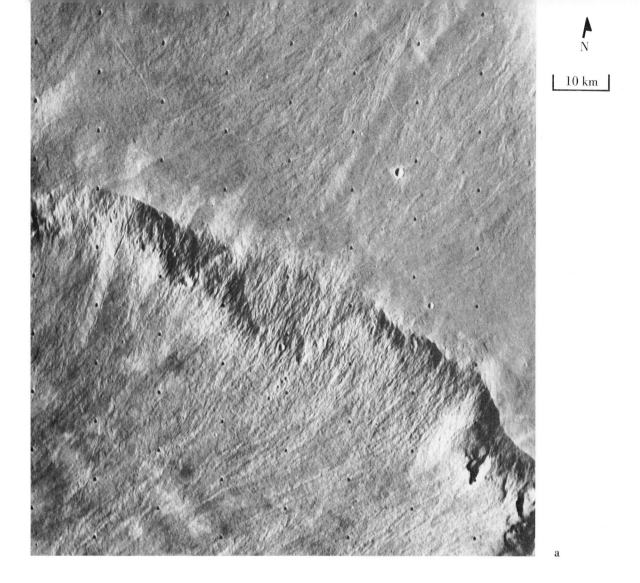

Lava Flow Drapes over Olympus Mons Scarp. (a) Lava channels and partially collapsed lava tubes are visible along the crests of ridge-like flows. The surface features on these flows are similar to those developed on basaltic flows on Earth. Clearly, the scarp in this area is older than the flows, indicating that at least the youngest flows on the mountain occurred after scarp formation. In this region, the Olympus Mons flows make up the plains surface at the foot of the scarp. However, in other areas, Olympus Mons flows have been overlain by the smooth-surfaced material of the plains. (b) The stereographic pair graphically portrays the ruggedness of the scarp. [(a) 47B25, (b) Left 46B34, Right 45B45; 21° N, 130° W]

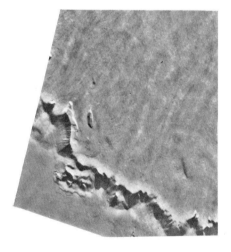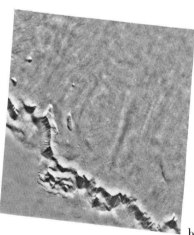

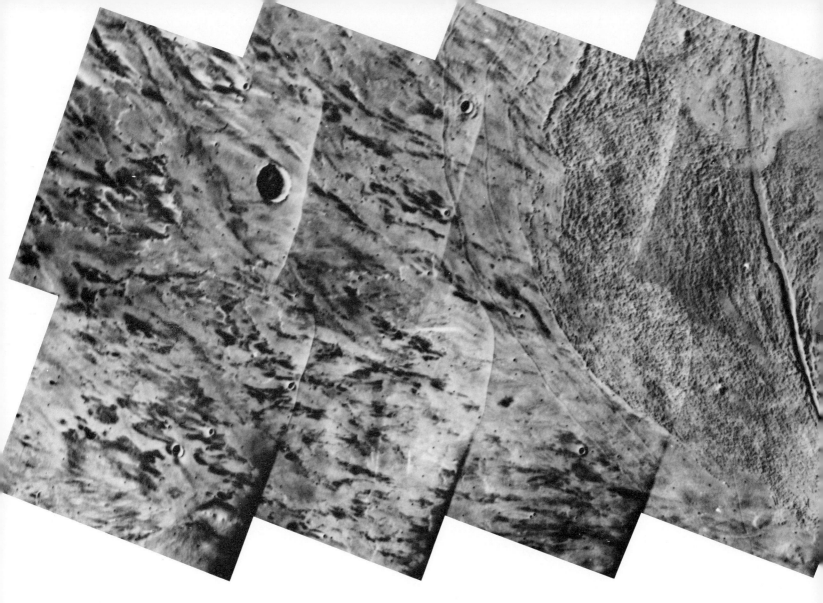

Arsia Mons. The summit is at about the same elevation as that of Olympus Mons, rising 16 km above the Tharsis Ridge—itself about 11 km high. The caldera is less complex than that of Olympus Mons, being a single, large circular structure about 140 km in diameter. Surrounding the caldera are concentric graben; the main northeast-southwest trending fracture zone underlying the volcano is indicated by numerous collapse pits seen here on the upper side of the caldera. This mosaic shows an enormous flow-like feature that extends from the volcano flanks onto the adjacent plains, and which consists of hummocky terrain with faint concentric features. The flow terminates in fine scale ridges parallel to the flow's front edge. The origin of this feature is not clear, but it may be a major landslide that developed high on the flanks of the volcano at a time when the volcano slopes were unstable. The concentric ridges in the distal parts appear to run through all the topographic features without substantially modifying them, and may be pressure ridges that developed in the underlying terrain at the foot of the unit. [211-5317; 9° S, 123° W]

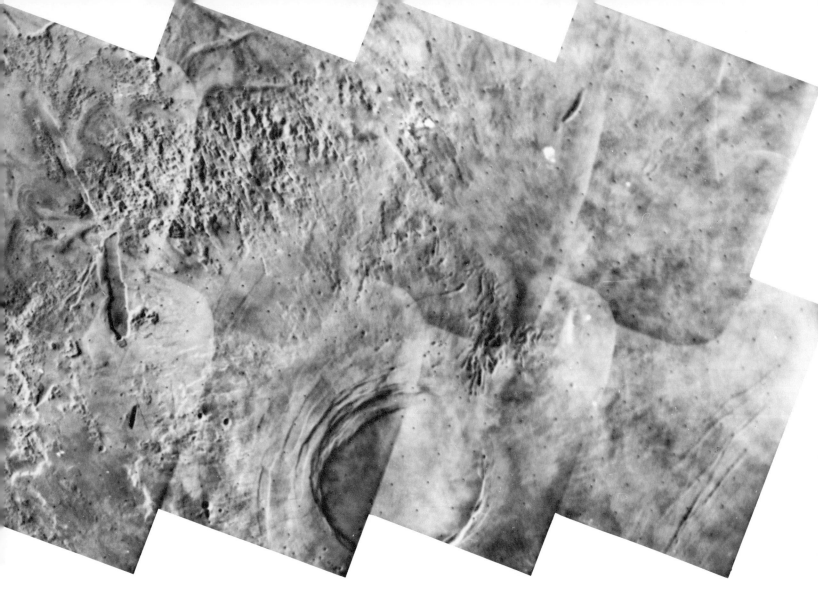

N

| 20 km |

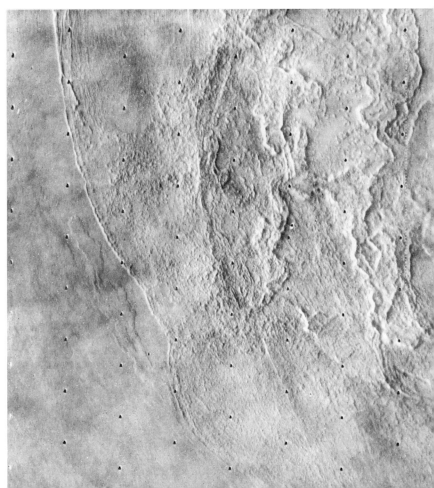

Possible Landslide Deposit on Arsia Mons. Hummocky terrain makes up most of this flow, and grades into the finely ridged, concentric flow front. These features may be pressure ridges at the front of the flow or, in some places, deceleration ridges formed as the flow came to a standstill. Small lava flow fronts are visible on the smooth plains in front of the main flow. [49B89; 3° N, 117° W]

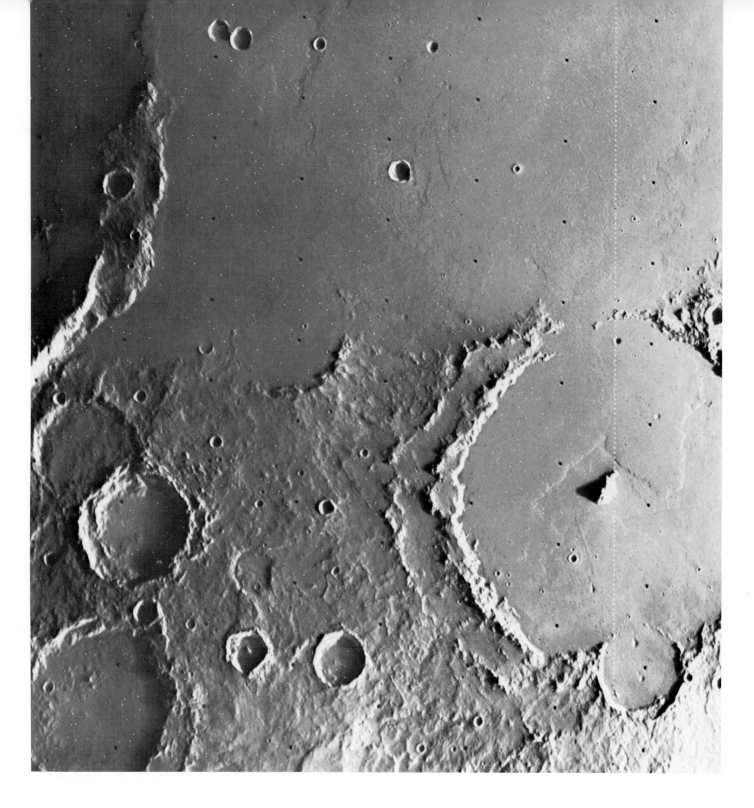

N

20 km

Extensive Lava Flows from Arsia Mons. The flows that erupted from Arsia Mons extend some 1500 km away from the summit and bury the older cratered terrain of the southern hemisphere. Flow fronts are visible within the large crater Pickering (120-km-diameter), where they have been diverted around high ground associated with the central peak of the crater. Flows of this type associated with the big volcanoes may have lengths in excess of 1000 km, and may resemble the large flows found in Mare Imbrium on the Moon. The discovery of these flows on the outer flanks of the major volcanoes on Mars has shown that the basal diameter of many of these volcanoes is considerably larger than was suspected from Mariner 9 data. [56A12; 34° S, 133° W]

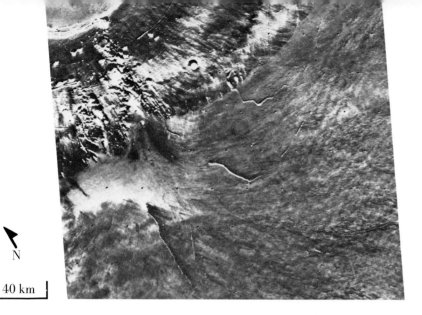

40 km

Arsia Mons Summit. Part of the caldera is visible at the upper left of the picture. The summit of the volcano is cut by lines of pits marking the fracture zone running through the volcano. Most of the lava at the middle and bottom right of the picture appears to have originated from the fracture zone, and postdates the summit cone of the volcano. A well-defined channel/tube system is visible toward the lower right of the picture; small pits at the head of this channel system represent the vent area. [52A04; 12° S, 120° W]

Summit of Alba Patera. This volcano is only a few kilometers above the surrounding plain which, coupled with its large diameter of some 1700 km, gives it a much lower profile than the Tharsis volcanoes. The rim of an old caldera near the summit, partly buried by younger lava flows, is visible at the bottom left; at the bottom right a younger caldera is at the top of the youngest summit cone. Lava flows are well preserved, and flows can be seen extending from near the lower right of the picture toward the upper left. [7B94; 41° N, 109° W]

20 km

N

|— 10 km —|

Lava-Covered Upper Flanks of Alba Patera. Different kinds of flows are visible. Large, relatively flat-topped flows with well-defined flow fronts occur in the middle of the frame. At the lower left are long flow-ridges, some of which extend for several hundred kilometers. The flat-topped flows are generally considered to have been fed by lava tubes. One flow has a sinuous channel-tube running along the crest of the ridge. Superposed impact craters on Alba Patera are more numerous than on Olympus Mons and Arsia Mons, suggesting an older age for many of these flows. [7B24; 48° N, 115° W]

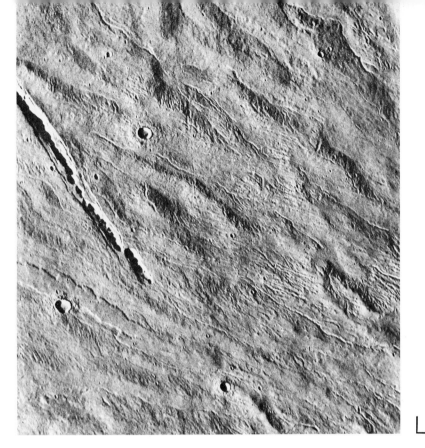

Ridge-Like Lava Flows on Alba Patera. This part of the flanks of Alba Patera has ridge-like lava flows with complicated dendritic patterns developed on them. Some of these channels may be directly associated with the formation of the lava flows, but some may have resulted from fluvial modification of the volcano flanks. Cutting the lava flows in this area is a well-defined graben, within which are numerous collapse pits. [7B53; 46° N, 119° W]

Biblis Patera. This volcano, situated between Arsia Mons and Olympus Mons, is much smaller than those so far described. Flow features on the flanks of the volcano are truncated by the surrounding plains, indicating partial burial by later deposits. The exposed part of the volcano has a basal diameter of about 100 km. Its original size may have been larger, although, from the small size of the caldera, it is unlikely—even considering the buried base—that it was ever as large as the giant Tharsis volcanoes. The summit caldera is surrounded by almost circular faults, which seems characteristic of Martian volcanoes. [44B50; 3° N, 124° W]

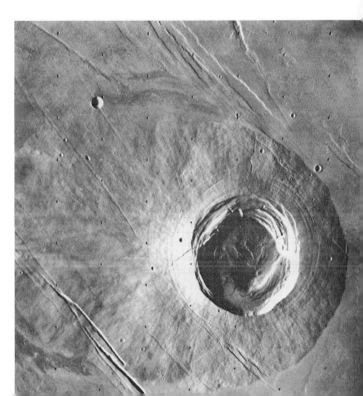

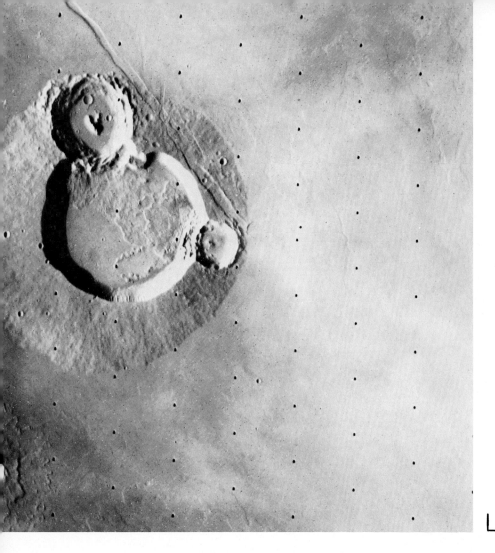

Ulysses Patera. This volcano lies just to the east of Biblis Patera in the northwest part of Tharsis. It is similar in size to Biblis Patera, is surrounded by younger flows, and has two superposed craters, probably of impact origin. These craters are older than the surrounding plains, and they have intersected the caldera walls and pushed material into the floor of the caldera. [49B85; 3° N, 121° W]

Tharsis Tholus. This 170-km diameter volcano differs in form from the volcanoes previously illustrated. The caldera has a wide bench around one side. This bench may represent an early lava lake level before further collapse occurred in the middle of the caldera. Scarps intersecting the caldera appear to be normal faults rather than graben. The base of the volcano is covered by younger materials so its original size cannot be determined. [225A13; 13° N, 92° W]

Tyrrhena Patera. The flanks of this ancient, southern hemisphere volcano have been strongly modified and embayed. At the summit is an irregular depression that is continuous with a valley, extending down the outer flanks. Concentric graben surround the summit. The volcano is so degraded that there are no well-defined primary volcanic depositional features to provide clues regarding the nature of the erupted materials. However, the low profile of the volcano, and the way in which outliers of the volcano form mesa-like bodies, suggest ash flow deposits rather than lavas. [211-5730; 20° S, 252° W]

Hadriaca Patera. This volcano's caldera is much better defined than that of Tyrrhena Patera, but its flanks are strongly degraded by radial valleys. The volcano is younger than many of the surrounding craters, but still much older than the Tharsis volcanoes, as indicated by the numbers of superposed impact craters. [97A42; 30° S, 270° W]

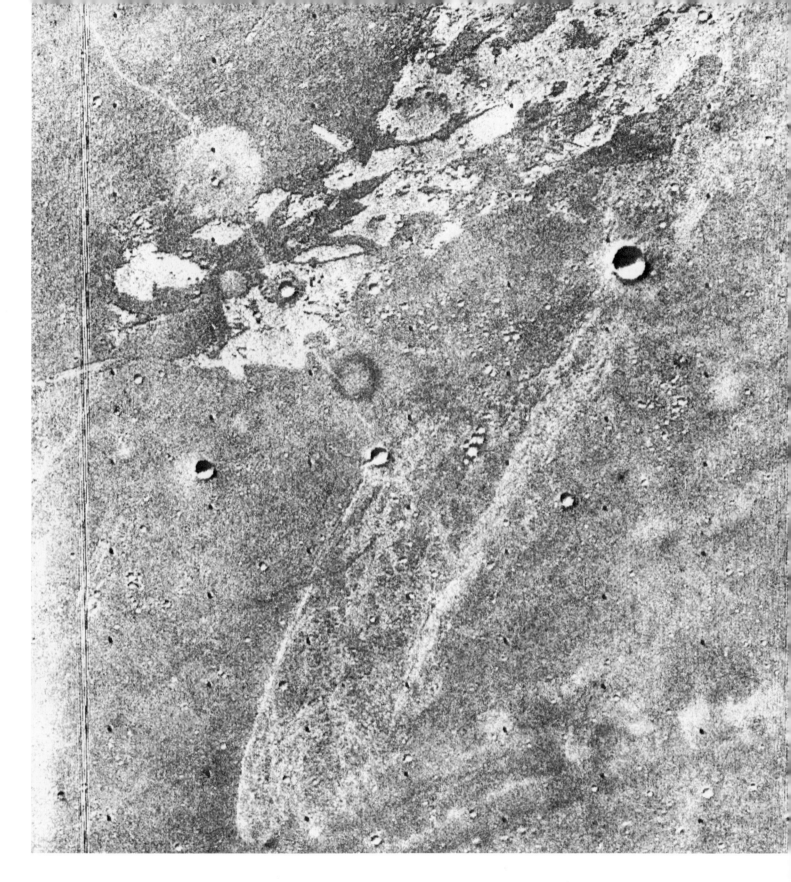

Volcano-Like Features in the Chryse Basin. In the upper left of this picture, a light circular feature with a central pit and a very low profile straddles a sinuous line, which may be the trace of a dike or fracture. The feature is less than 10 km across, much smaller than any of the other volcanoes described. [4A36; 18° N, 35° W]

DEFORMATIONAL FEATURES

ALTHOUGH DEFORMATIONAL FEATURES are common on the Martian surface, the type of deformation differs from that on Earth. Deformation of Earth's surface is controlled largely by plate motion. Where plates converge, intense folding, overthrusting, and transcurrent faulting result, and mountain chains may form. Where plates diverge, as at midoceanic ridges, tensional features develop, but they are commonly masked by volcanic deposits. Apparently no plate motion occurred on Mars, and the deformational features associated with plate motion on Earth are absent. The dominant type of deformation on Mars is normal faulting; compressional and transcurrent features, although present, are rare.

Most faults are associated with the Tharsis uplift, a 6000-km diameter, 7-km high bulge in the Martian crust roughly centered on Tharsis. Around the periphery of the bulge, and aligned approximately radial to it, are numerous fractures, some of which extend as far as 4000 km from the center. So extensive are the radial fractures that they are the dominant structural feature of the entire hemisphere. Fracturing seems to vary greatly in intensity and age. In some places, such as in the Ceraunius and Tantalus Fossae (north of Tharsis) and the Claritas Fossae (south of Tharsis), fracturing is extremely intense; other areas are completely free of fractures. The number of craters superposed on the fractures is a measure of their relative ages and indicates a wide range of ages. Thus, fracturing associated with the Tharsis bulge evidently has continued for much of the planet's history.

Although fractures around Tharsis include the most prominent tectonic features on the planet, several fracture systems seem unrelated to Tharsis. Some fractures occur around old impact basins and are generally concentric to them. Especially prominent are the Nilae Fossae around the Isidis basin, but less distinct concentric graben and scarps are visible around the Argyre and Hellas basins. Dominantly northeast-southwest and northwest-southeast lineaments are detectable throughout much of the old cratered terrain as escarpments or linear sections of crater walls. Where the old cratered terrain is eroded, as in the fretted terrain, the erosion occurred preferentially along these directions.

Mare ridges are other possible examples of deformational features. Such ridges are common on the sparsely cratered plains of Lunae Planum, Syria Planum, Hesperia Planum, and around the site of Viking Lander 1 in Chryse Planitia. In fact, Viking Lander 1 is believed to have landed on a ridge crest. Similar ridges have been studied intensively on the Moon and are considered to be the surface expression of reverse or thrust faults, which formed either contemporaneously with deposition of the lunar mare rocks, or some time after, as a result of accommodation of the Moon's surface to the lava deposition. The ridges on Mars have a striking resemblance to those on the Moon and are probably of similar origin. Terrestrial analogs have not been found, however, and their origin remains uncertain.

Most of the cratered plains in the northern latitudes of Mars exhibit a polygonal pattern of fractures for which there is no terrestrial analog. Individual polygons average approximately 10 km across and extend uniformly in all directions. Ice wedging and contraction by cooling have been suggested as possibilities, but no completely satisfactory explanation has yet been found.

Ceraunius Fossae. Sparsely cratered plains seen here have been intensely fractured by closely spaced parallel faults. Individual faults can be traced for up to 300 km. The faults are part of the system of fractures radial to the Tharsis bulge. [39B59; 25° N, 101° W]

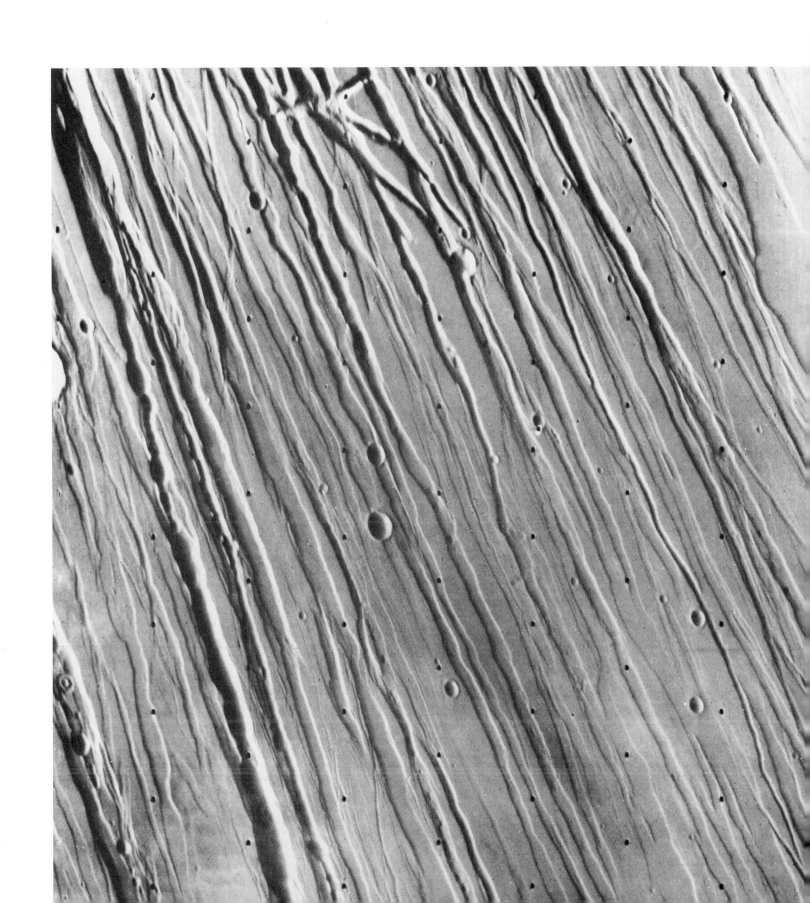

Chains of Rimless Pits within Graben of Ceraunius Fossae. Rimless depressions commonly occur in the graben of this area. The pits do not seem to be sources of the extensive lava flows visible in the picture, but instead cut across flows and some fractures. The lines of pits are usually located within graben and not on the intervening plains. [224A13; 32° N, 102° W]

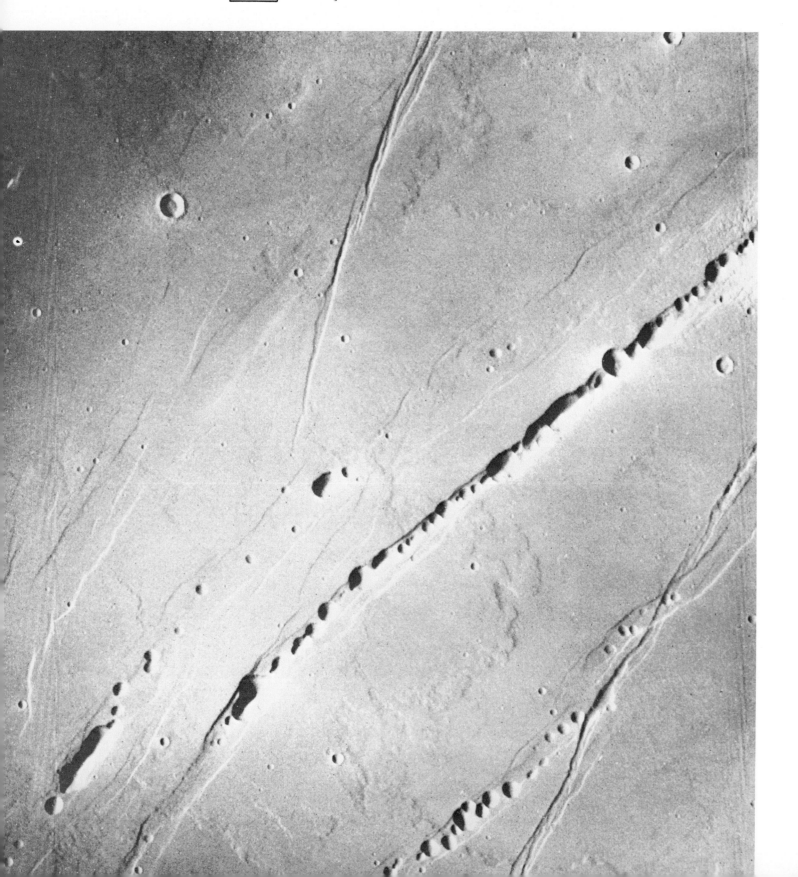

N

50 km

Fractured Terrain of the Thaumasia Region. An escarpment in the center of this picture is at the south extension of the end of Claritas Fossae. The fractures are roughly radial to the Tharsis bulge and cut mostly old cratered terrain. Crater counts indicate that most of the fractures are older than the corresponding fractures north of the Tharsis bulge. [57A04-13; 37° S, 103° W]

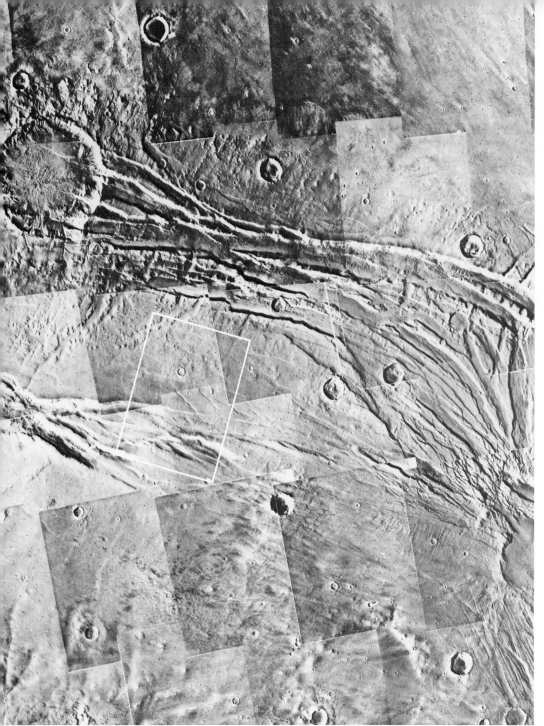

Fractured Terrain North of Olympus Mons. (a) This fracture is part of an old arcuate structure that is partly buried to the south by lavas from Olympus Mons. (b) This stereogram shows the area outlined in (a), and gives a greater sense of the steepness of the walls in the fractures. [211-5528; 38° N, 131° W]

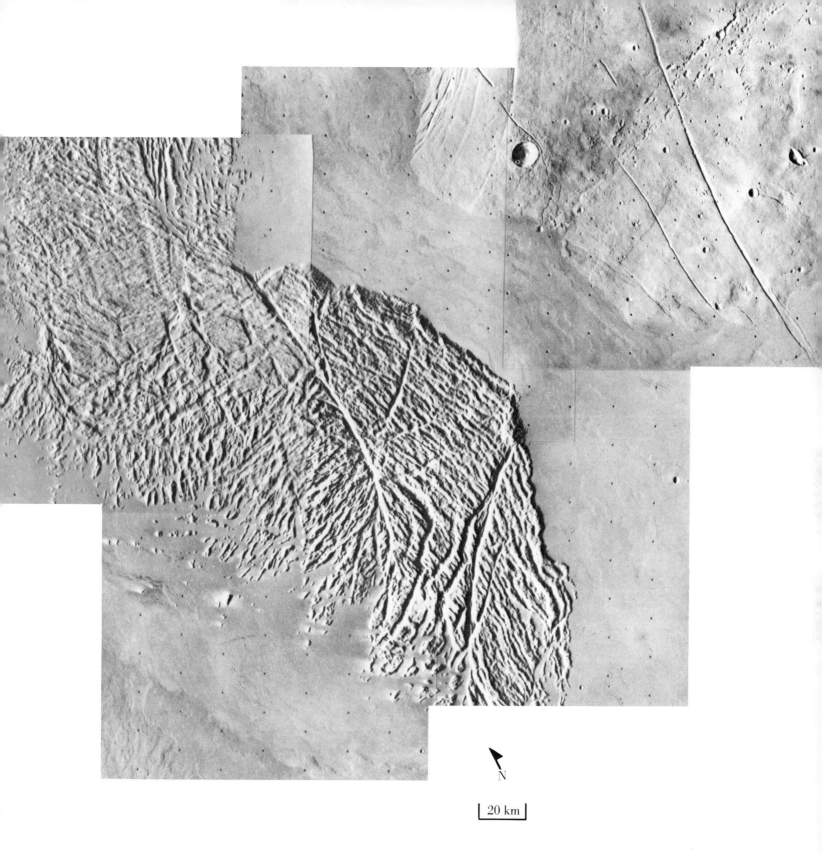

Grooved Terrain around Olympus Mons. Grooved terrain, the term applied to the fractured surface in the bottom left quadrant of this picture, occurs in a discontinuous ring around Olympus Mons. The origin of the terrain is unknown, but it has been suggested that it occurs at the surface of vast thrust sheets caused by the loading of the crust by Olympus Mons. Another suggestion is that the terrain is formed by erosion of ash flow tuffs that originated from Olympus Mons. To the northeast, young lava flows transect the grooved terrain and an older fractured surface. [48B43-47; 32° N, 132° W]

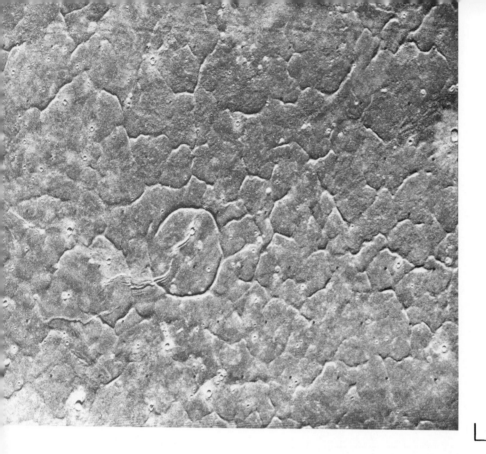

10 km

Fractured Plains. Large areas of plains in the 40° to 50° N latitude belt have a fracture pattern similar to that shown here. The pattern resembles those that form on some lava lakes as a result of cooling. It also resembles patterned ground formed by ice-wedging in periglacial regions. The polygons on Mars are, however, approximately 100 times the size of the suggested terrestrial analogs, and their origin is unknown. [32A18; 44° N, 18° W]

Fractured Plains. The fracture pattern here is coarser than that shown in the previous photograph. Most of the crevasses have flat floors. The low hills at the bottom of the picture may be erosional remnants of old cratered terrain, the main body of which occurs just to the south. [35A64; 40° N, 14° W]

10 km

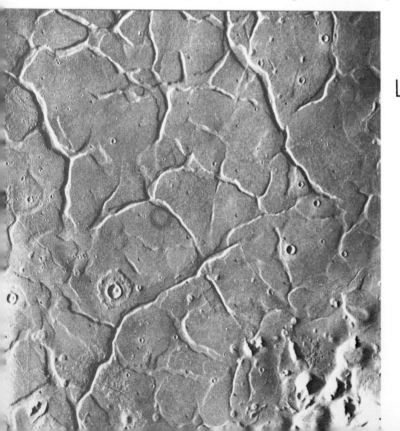

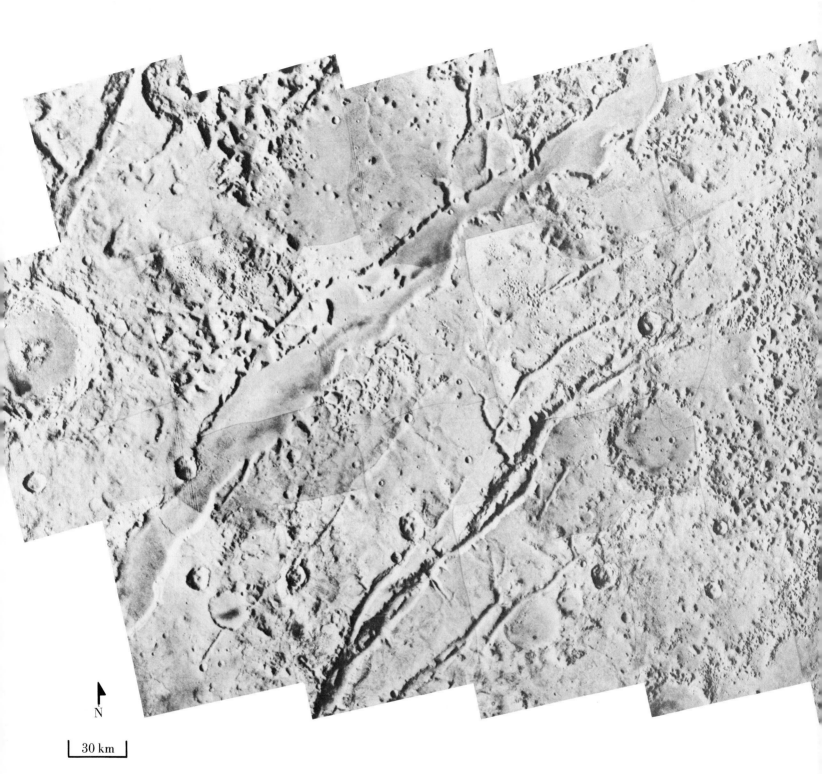

Nilae Fossae. The fractures seen here are concentric to what is probably an old impact basin centered on Isidis Planitia. The fractures appear to be very old, as indicated by the superposition of large impact craters. [211-5657; 25° N, 282° W]

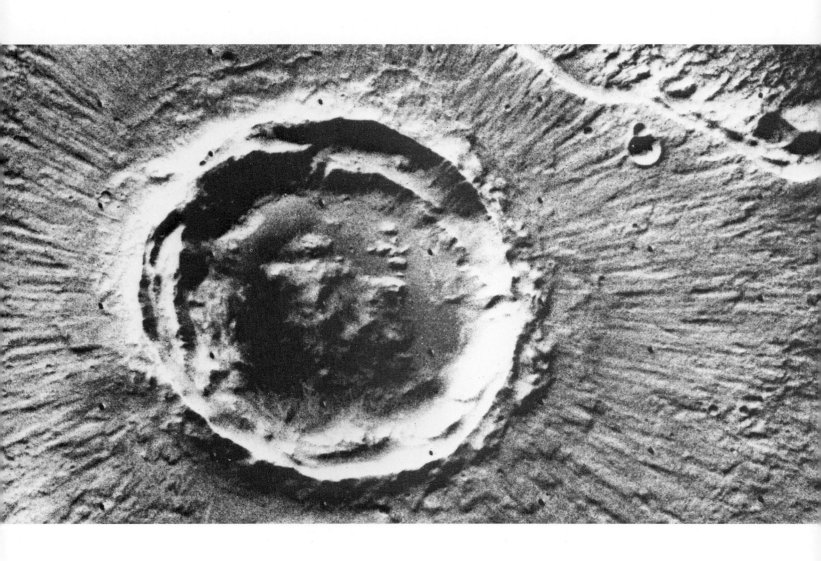

CRATERS

IMPACT CRATERS, which occur almost everywhere on the Martian surface, are significant because the number of impact craters per unit area area gives an indication of the relative ages of different parts of the surface. They also provide clues to the properties of the near-surface materials and record the effects of various processes, such as wind action, that modify the surface. The density of impact craters varies from the heavily cratered southern hemisphere to very sparsely cratered regions like the polar dune fields and laminated terrain.

Most surfaces photographed by the Viking orbiters have crater densities in excess of typical lunar mare surfaces. This condition suggests that most of the Martian surface is probably billions of years old. The implication is that resurfacing, such as by volcanic processes or wind action, is extremely slow in most places compared with Earth. Crater densities low enough to suggest significantly higher rates of resurfacing are found only in the polar regions, around some volcanoes, and very locally in other areas. In the 40° to 60° latitude belts, particularly, significant surface modification has occurred since the presently observed crater population formed. Craters in these areas have apparently been modified by repeated burial and stripping of debris layers. The era in which this activity occurred, and whether or not it is continuing, is unknown.

The Viking orbiter pictures revealed some unique characteristics of Martian impact craters. The ejecta pattern around most fresh Martian impact craters is distinctively different from that around lunar and mercurian craters. On the Moon and on Mercury, the ejecta typically have a coarse, disordered texture close to the rim. Farther out, the texture becomes finer and, with increasing radial distance, grades imperceptibly into dense fields of secondary craters and rays. Most Martian craters have a quite different ejecta pattern. The ejecta commonly appear to consist of several layers, with the outer edge of each marked by a low ridge or escarpment. Features on the ejecta surfaces include closely spaced radial striae and concentric grooves, ridges, and scarps, especially toward the outer margin. These unique Martian features were seen vaguely in the Mariner 9 pictures and tentatively attributed to wind action.

Viking pictures show that many of the peculiar characteristics of Martian craters are primary emplacement features not due to wind. The fresher the crater appears, the better preserved are the striae, ramparts, and concentric features. Very small secondary craters indicate that the crater has undergone very little modification since its formation. Martian craters look different from those on the Moon and Mercury because the process of ejecta emplacement is different. The final stage of emplacement of ejecta on Mars is thought to be an outward moving debris flow instead of the simple ballistic deposition that occurs on the Moon and Mercury. Why the emplacement process is so different on Mars is not understood; possibly encountering water at depth or the melting of large amounts of ground ice by the impact results in fluid, mud-like ejecta. The presence of an atmosphere may also have an effect on the manner in which the ejecta are emplaced. These characteristics of the craters indicate that the properties of near-surface materials on Mars are quite different from those on the Moon and Mercury.

Although some features of Martian craters formerly attributed to wind action are now believed to be primary, modification by wind is still significant in many areas. In the 40° to 60° N latitude band, numerous craters that have been termed pedestal craters occur in the center of a roughly circular platform many crater-diameters across. The pedestals have been attributed to the partial stripping of a formerly continuous debris blanket by the wind. The general surface layer of debris has been removed everywhere except around craters where ejecta have armored the surface against the wind, with the result that most craters are surrounded by a platform composed of remnants of the debris blanket. This feature is especially striking where fields of secondary craters occur, and results in arrays of low hills with central craters that simulate a field of volcanic cones. In other areas, repeated burial and stripping have led to bizarre formations, such as ejecta surfaces at a lower elevation than the surrounding terrain.

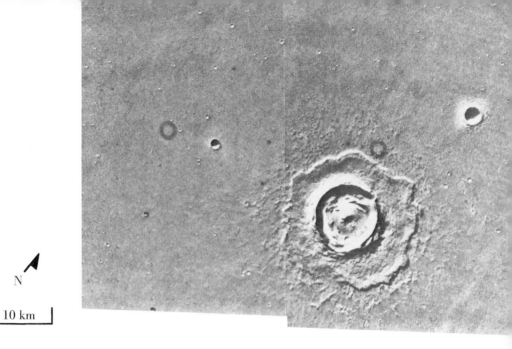

Rampart Crater in Chryse Planitia. The outer edge of the inner ejecta layer of Belz Crater is demarcated by a low ridge or rampart. On the surface of the ejecta layer are faint radial striations and concentric ridges and grooves. Outside the rampart, the topography is similar to the distal parts of lunar and mercurian ejecta, with numerous isolated hummocks and indistinct radial ridges. [10A54, 56; 22° N, 43° W]

Layered Ejecta around the Crater Tarsus. Each ejecta layer seen here has an outer ridge or escarpment. The upper layer appears to have flowed over and transected the outer margin of the lower layers. At the arrow, ejecta have flowed around a low obstacle. [10A66, 68, 70, 92-98; 23° N, 40° W]

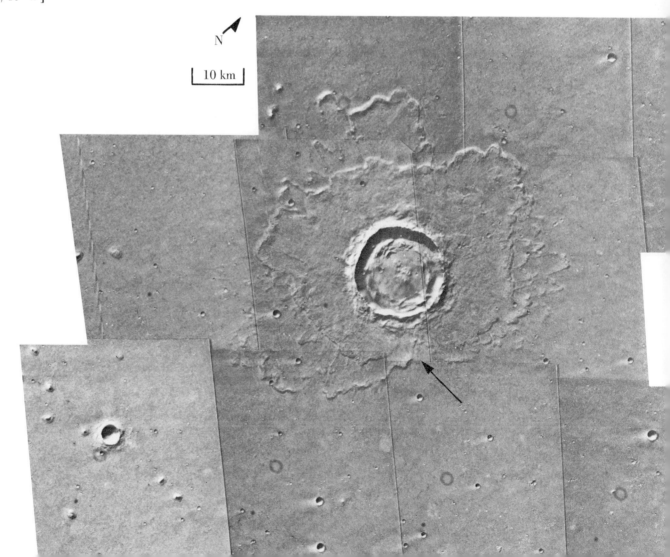

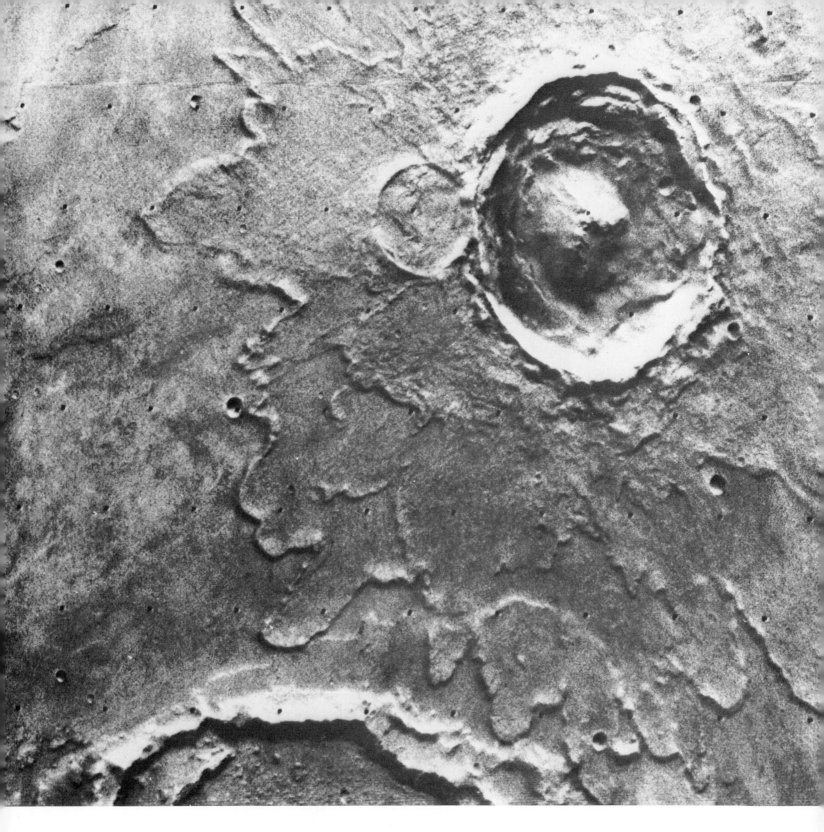

Yuty Crater. This crater has several ejecta layers, each complexly lobed and each with an outer rampart. Although buried by Yuty ejecta, a pre-Yuty crater close to the rim is clearly visible, indicating that the ejecta deposit is thin. [3A07; 22° N, 34° W]

Arandas Crater. The outer layer of ejecta has flowed over the surrounding fractured plains. The two arrows indicate where an inner ejecta layer has flowed around pre-existing craters. Numerous low ridges occur on the inner ejecta layer close to and parallel to its outer margin. [32A28-31, 9A42; 43° N, 14° W]

Detail of Arandas Crater Ejecta. The fractures of the underlying plains at Arandas' location are clearly visible through the ejecta, even close to the rim, showing that the ejecta are very thin. On the ejecta surface is a fine radial pattern. [32A28; 43° N, 14° W]

Poona Crater. This crater is close to Kasei Vallis, the edge of which is marked by an escarpment in the northwest corner of the image. The ejecta have a marked radial pattern and no outer rampart. [22A54; 24° N, 52° W]

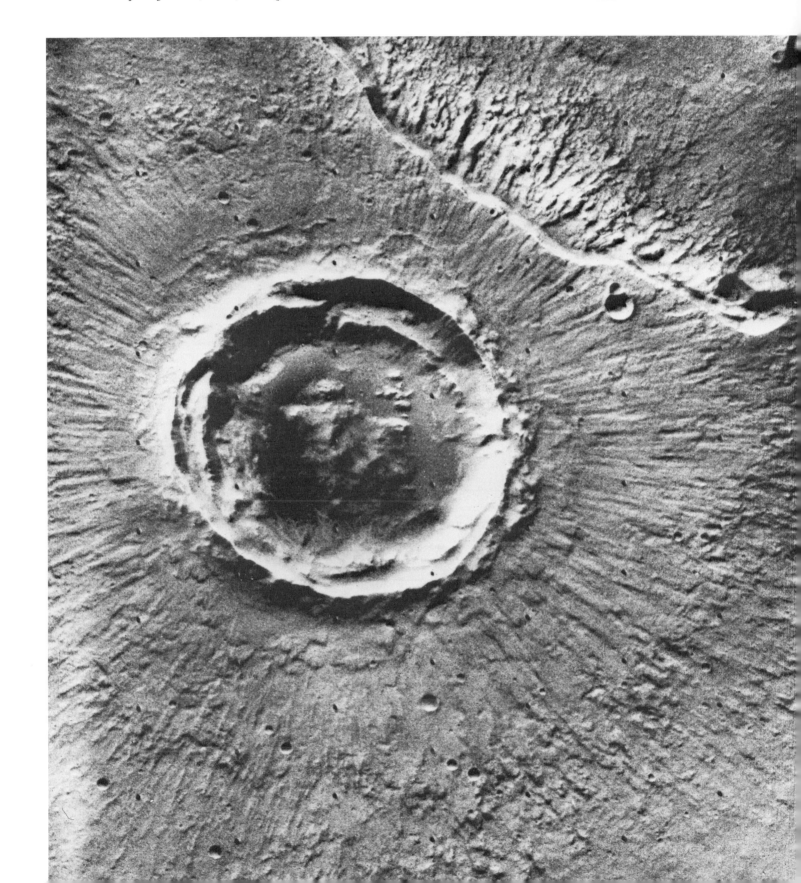

Pedestal Craters. Almost all the craters in this area are situated within a pedestal or platform that stands above the surrounding plains. The diameter of the platform decreases in size as the crater diameter decreases, so that small craters may occur atop low domes. The mode of formation for the pedestal craters is poorly understood. The observed configuration may be partly primary and partly the result of selective stripping of a former layer of debris that covered the surface, with the layer now remaining only around craters. [43A04; 46° N, 353° W]

Pedestal Craters. The ejecta deposit around the crater in the left half of the frame still retains its multilayered character and faint radial surface texture. Other craters have well-developed pedestals, but primary textures are less well preserved. A cluster of craters at the top of the frame strongly resembles a group of volcanic cones but is probably part of the impact crater continuum. [60A53; 48° N, 349° W]

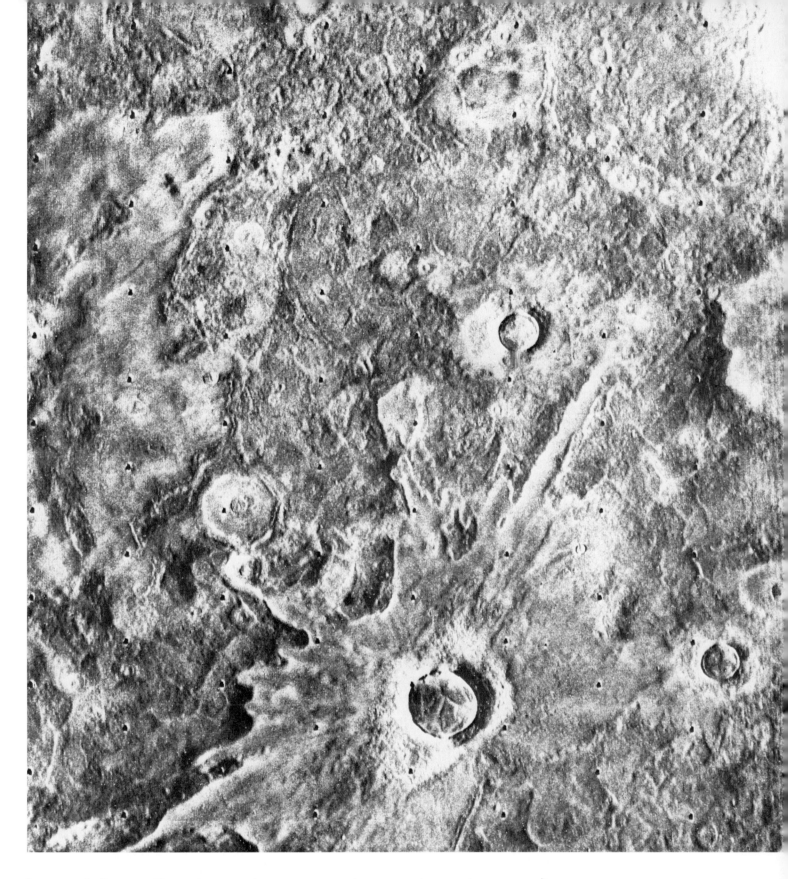

Craters with Irregular Ejecta Blanket. Some craters in this picture have well-developed pedestals; others are surrounded by bright ejecta but no well-defined outer escarpment. The large crater in the bottom half of the frame has a highly irregular ejecta pattern with elongate lobes extending out over the surrounding fractured plains. [10B52; 45° N, 259° W]

Lowell Crater. As on the Moon and Mercury, very large impacts form multiringed basins such as this 210-km diameter crater Radial lines of secondary craters are clearly visible outside the more disordered terrain close to the rim crest, despite partial dissection by small channels. The smooth feature in the center of the basin may be a low-lying cloud. To the north of Lowell, near the center of the picture is Slipher Crater. [34A17-20; 52° S, 81° W]

 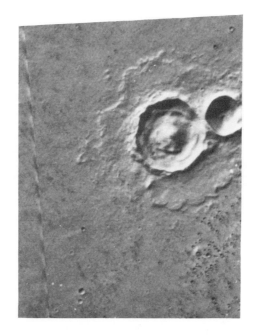

Hamelin Crater. Hamelin Crater, only 90 km northeast of the original landing site selected for Viking Lander 1 in Chryse Planitia, shows the raised edge or rampart around the ejecta blanket that is characteristic of many craters in the area. [77/08/10/180701; 20° N, 33° W]

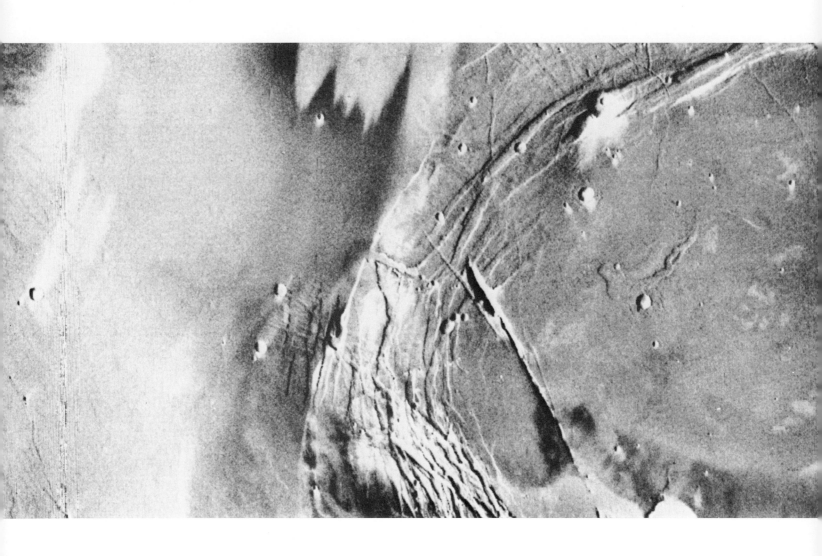

VARIABLE FEATURES

ALBEDO VARIATIONS on the surface of Mars can be attributed to aeolian phenomena or to the deposition and sublimation of volatiles in the polar regions. No evidence exists for the seasonal wave of darkening once proposed by some observers.

Very few surface changes were noted during the Viking primary mission in 1976, a finding consistent with the predictions of relatively low wind velocities during northern summer. However, with the approach of southern summer, wind activity increased, and many albedo changes were noticed that were similar (but not identical) to those recorded by Mariner 9 in 1972 at a comparable season.

Comparisons of specific albedo boundaries in the Mariner 9 and Viking pictures showed that, in many cases, subtle changes in outline and/or contrast had occurred; in a few areas, e.g., Syria Planum, the albedo boundaries were dramatically different.

The most conspicuous wind markers on the planet are light, crater-associated streaks whose pattern is that of the wind flow expected during southern summer (when surface winds are strongest and major dust storms occur). Although some new light streaks formed between 1972 and 1976, and a few old ones disappeared, most light streaks were essentially unchanged in outline and direction. These bright streaks probably are deposits of dust, as storm fallout, which accumulates in the lee of craters and other topographic obstructions to the wind flow.

Viking observations confirm that the average lifetime of ragged dark streaks, usually interpreted as erosion scars, is shorter than that of the light streaks. Many dark streaks changed conspicuously in both outline and direction, not only between 1972 and 1976, but also during the Viking mission itself.

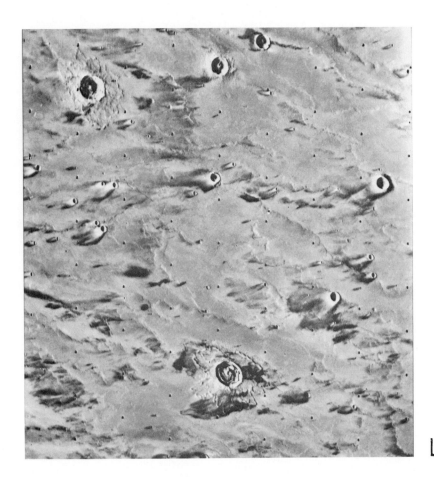

40 km

Mixed-Tone Streaks in Memnonia. These distinctive mixed-tone streaks appear to consist of a central, tapered light streak bordered by two dark side-lobes. Similar streaks have been produced in wind tunnel simulations. If the wind tunnel results are valid, the dark lobes represent areas of wind erosion, and the bright central portion is a region of deposition where bright fine particles accumulate. Numerous ragged dark streaks are also visible, and are interpreted as zones of erosion associated with topographic obstacles—in this case including crater ramparts, ejecta blankets, and possible lava flow fronts. [41B51; 13° S, 139° W]

▷

Mesogaea Area. Only subtle albedo changes have occurred in this complex mixed-tone streak since the Mariner 9 coverage. The complex streak seems to be the result of deflation of low albedo material from the crater at upper right. The numerous bright streaks outside the large mixed-tone streak are interpreted as accumulations of dust storm fallout. Craters and isolated hills seem to produce similar bright streaks. Bright areas within the main dark streaks could be either deposits of dust storm fallout or shadow zones behind topographic obstacles where dark material from the upwind crater was not deposited. Thus, the conspicuous bright, hill-associated streak within the main dark streak could be a normal bright streak or a "shadow" streak. [88A81-88; 8° N, 192° W]

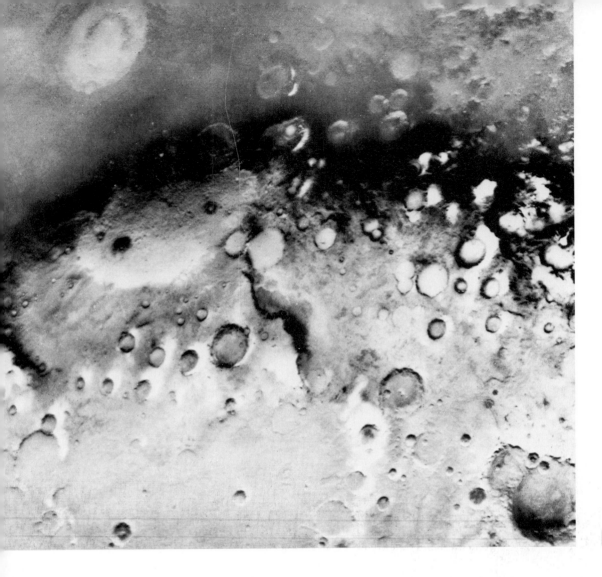

Frost Streaks in the Annual South Polar Cap. These bright, streamlined albedo features are associated with craters near the retreating margin of the annual frost cap. The features become more prominent (relative to the background) as the cap edge approaches their location, and disappear shortly after the margin passes. The bright streaks are interpreted as accumulations of carbon dioxide frost in the lee of craters, and suggest that winds may be effective in redistributing frost in the polar regions of Mars. The wind direction indicated by the streaks suggests that the streaks are laid down during southern winter. [161B26; 61° S, 71° W]

Contrast Reversal in the Cerberus Region. These frames show the contrast reversal of crater-and-hill associated streaks. Through a red filter (left, about 0.58 μm), the streaks have a "normal" appearance and are brighter than their surroundings; through a violet filter (right, about 0.45 μm), they appear darker. Laboratory measurements indicate that such contrast reversal is a common property of well-sorted, very fine samples of iron oxide materials. Not all bright streaks on Mars show such contrast reversal in the violet. [Left 53B56, Right 53B57; 12° N, 202° W]

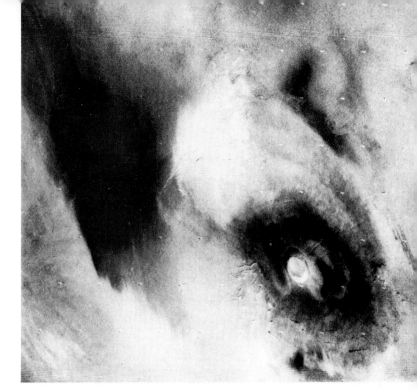

a   b

Aeolian Activity on Pavonis Mons. Persistent aeolian activity was observed on the flanks of the three large Tharsis volcanoes. Early Viking imaging revealed that since the last Mariner 9 coverage in 1972, each of the three large Tharsis volcanoes (Ascraeus, Pavonis, and Arsia), had developed a more or less complete dark albedo ring on its flanks. The dark albedo ring was especially well-developed on Pavonis Mons, as seen here, where it was 20 km wide and situated at altitudes between 20 and 25 km.

(a) The boundaries of the dark albedo ring are ragged, and the upper boundary is composed partly of coalescing, ragged dark streaks trending downhill. (b) Another view shows the same area taken after the 1977 global dust storm. The observed changes are best explained by the erosion by downslope winds of bright albedo material—probably storm fallout. Aeolian activity has been observed up to the summits of the Tharsis volcanoes, proving that Martian winds are strong enough to transport fine particles even at the very low pressures at the tops of the Tharsis volcanoes. [Left 52A15, Right 416A45; 0° N, 113° W]

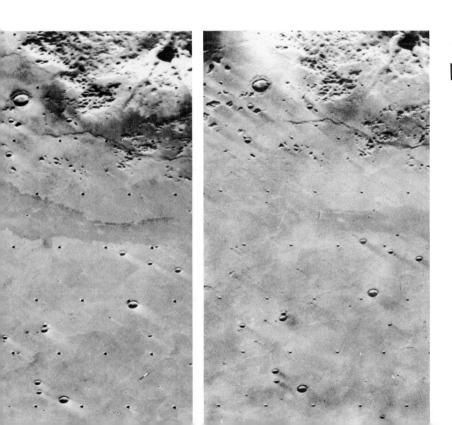

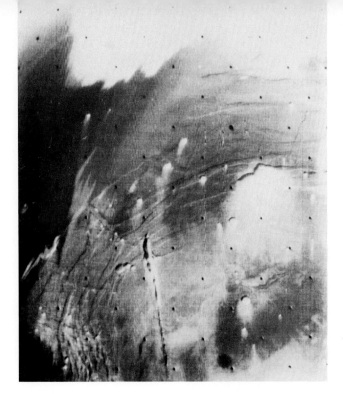

a

Spectacular Albedo Changes in Syria Planum. These four views of Syria Planum show both long-term and short-term changes in the surface markings. (a) A Mariner 9 view of the area in southern summer is shown. (b) A Viking Orbiter 1 view of the same area almost three Martian years later was taken shortly after the start of the global-scale storms. (c) and (d) These were taken in mid to late southern summer. The changes observed in the bright streaks in the last three frames are attributed to strong north-to-south winds during the global dust storms. [(a) Mariner 9 DAS 08585544, (b) 294A69, (c) 416A49, (d) 439A48; 12° S, 110° W]

b

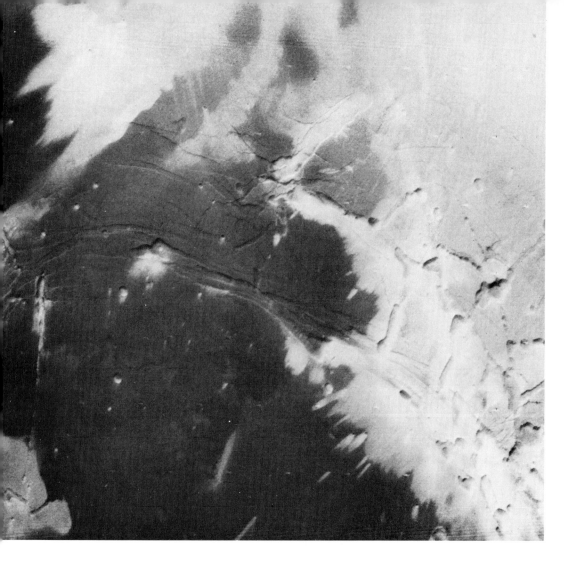

c

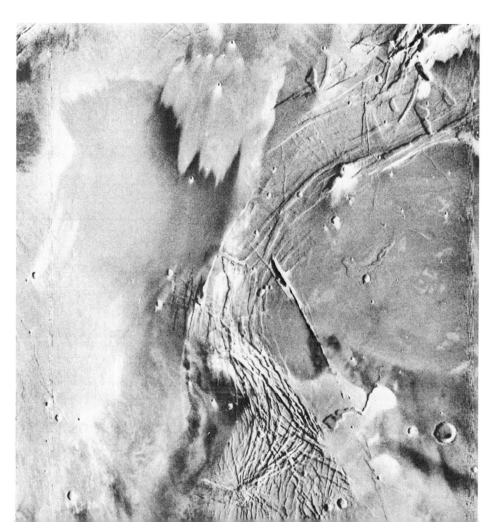

d

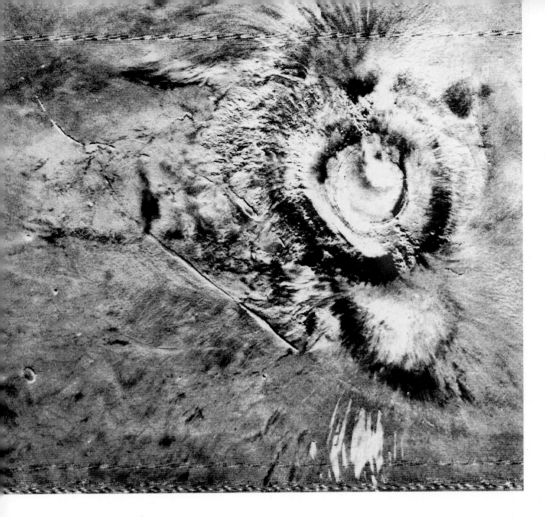

Changes in Wind Streaks on the Slopes of Arsia Mons. The changes in both dark and light streaks shown occurred after the global dust storms. In the lower image, winds blowing down the long slopes of the volcano redistributed some of the light coating of dust deposited during the global dust storms, forming both dark streaks (erosion of dust) and bright streaks (deposition of additional dust). [574A46, 648A03; 9° S, 124° W]

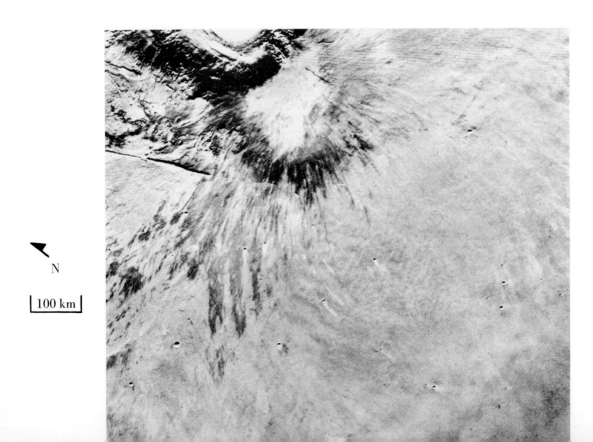

Erosion of Dust from Large Areas Following the Global Dust Storms. (a) Shortly after the global dust storms, wind erosion downwind of craters produced dark streaks; a combination of global circulation and local winds blowing down slopes formed streaks pointing in two directions from the same craters. (b) Later winds, also blowing downslope, stripped dust from much wider areas to form the large dark markings. [603A08, 639A67; 31° S, 117° W]

MARTIAN MOONS

TWO MOONS orbit Mars: Phobos (mean diameter, 22.0 km) and Deimos (mean diameter, 14.0 km). As part of the centennial celebration commemorating Asaph Hall's discovery of Phobos and Deimos in 1877, an extensive exploration of the two Martian moons was conducted with the Viking orbiters. The spectacular high-resolution imaging data obtained have rivaled in resolution any previous flyby or orbiter imaging data on any body in our solar system.

These data provided much more knowledge of the moons' surface morphology and their physical and dynamical properties. Phobos was observed to be somewhat smaller than determined by Mariner 9 ($\sim$5200 km^3 rather than 5700 km^3), and Deimos was somewhat larger ($\sim$1200 km^3 rather than 1000 km^3). Both satellites are locked into a stable, synchronous rotation about Mars, with their longest axes pointing toward Mars and their shortest axes normal to their orbit planes (which are within a few degrees of Mars' equator). Both satellites have topographic variations as large as 20 percent of local mean radii, and Deimos has a few large flat areas.

Viking found Phobos and Deimos to be within 10 to 15 km of their predicted positions based on Mariner 9 images. Precessing ellipses accurately model the orbits of the two moons, with short-period Mars gravity perturbations having displacement amplitudes of less than a few kilometers on Phobos' orbit, and solar perturbations having displacement amplitudes of less than 5 km for Deimos (except for one 110-km, 54-year periodic longitude perturbation).

Phobos, one of the three satellites in our solar system whose period ($7^h 39^m$) is less than the rotational period of the primary planet ($24^h 37^m$ for Mars), is losing orbital energy to surface tides it raises on Mars. As the orbit of Phobos decays and gets closer to Mars, Phobos may eventually be torn apart when the tidal forces of Mars overcome the cohesive bond between its particles. Phobos, already inside the "Roche Limit" where internal gravity alone is too weak to hold it together, could conceivably become a ring plane about Mars within the next 50 million years.

Phobos and Deimos are both uniformly gray. Albedos of $\sim$0.06 put both in a class with the darkest objects in our solar system. These dark surfaces appear to be layers of regolith with depths of a few hundred meters for Phobos and at least 5 to 10 meters for Deimos. Cratering of the surface of Phobos continued during and after the formation of the regolith, and the regolith is saturated with craters. However, on Deimos it appears that the regolith continued to develop after the cratering subsided, and the smaller craters (<100 meters) are partially filled or covered. This obscuration of the smaller craters gives Deimos a much smoother appearance than Phobos when viewed at ranges of more than a few hundred kilometers, because the filled craters are near or below the resolving power of the cameras and therefore are not visible.

In contrast to the smooth appearance of Deimos, the surface of Phobos is dominated by sharp, fresh-looking craters of all sizes and a vast network of linear features resembling crater chains. These linear grooves, up to tens of kilometers long and hundreds of meters across, appear to be surface fractures associated with the formation of Stickney, the largest crater on Phobos. Crater densities on both satellites are comparable to densities on the lunar uplands, a fact that suggests ages of up to a few billion years. However, impact fluxes may have been significantly higher for Phobos and Deimos because of ejecta being thrown into orbit about Mars and then recollected as the satellites swept it up in their orbits.

Similar networks of striations have not been identified on Deimos; however, they may have been covered by regolith, and picture resolution may not have been sufficient to identify such features. For example, a large depression 10 km across at the south pole of Deimos may have been caused by a single impact or may have been the result of fragmentation if Deimos was once part of a larger body. Linear features radiating from the center of this depression are suggested by the data, but low picture resolution has limited any interpretation of these features or determination of the origin of the large depression.

The close encounters with Phobos and Deimos have yielded preliminary mass determinations of approximately 1×10^{16} and 2×10^{15} kg, respectively. Using the volumes mentioned earlier, mean densities of about 1900 kg/m^3 for Phobos and 1400 kg/m^3 for Deimos are obtained. These low densities, as well as their colors and albedos, make Phobos and Deimos compositionally similar to Type-1 carbonaceous chondrites found in the asteroid belt. These data strongly suggest capture as the origin of the two asteroid-like moons of Mars.

Viking also obtained pictures of Phobos and Deimos, or their shadows, against Mars. The transit pictures were used in refining knowledge of the shapes of the satellites, and the shadow pictures helped locate Viking Lander 1. The satellite and shadow images were used to improve map coordinates of features on Mars surrounding the images.

▷

Phobos Closeup. The photomosaic on top was taken at a range of 300 km as Viking Orbiter 1 was approaching Phobos. The areas covered by three pictures taken at a range of 110 to 130 km are outlined on the photomosaic. The upper right area of the photomosaic shows a region dominated by grooves. The grooves are probably fractures in the surface of Phobos from a large impact. Two large craters with dark material on their floors are seen near the bottom of the photomosaic. These flat-bottomed craters give evidence that Phobos is covered by a regolith of up to a few hundred meters thick. The three pictures show the heavily cratered surface; craters as small as 10 to 15 meters are visible. [Top 211-5356, Left 244A03, Center 244A04, Right 244A06]

Phobos from 480 Kilometers. Viking Orbiter 1 flew within 480 km of Mars' inner satellite, Phobos, to obtain the pictures in this mosaic of the asteroid-size moon. As seen here, Phobos is nearly 75% illuminated and is about 21 km across and 19 km from top to bottom. Some features as small as 20 meters across can be seen. Surface features include grooves resembling linear chains of craters and small hummocks which appear to be resting on the surface. The regolith-covered surface is saturated with craters. Hummocks, mostly seen near the terminator (right), are about 50 meters in size and may be surface debris from impacts. [211-5353]

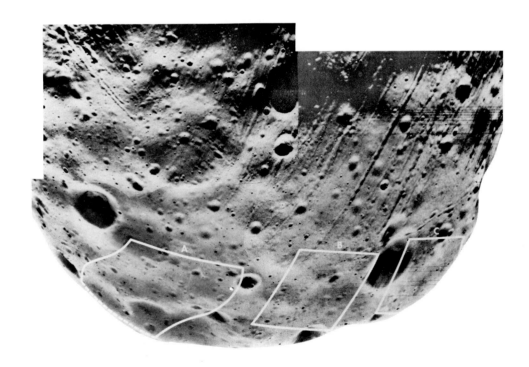

Fractures Radiating from Stickney Crater. Viking Orbiter 1 flew within 300 km of Phobos in May 1977 to obtain this photomosaic. Raw pictures are at the top and computer-enhanced pictures, to show small surface detail, are at the bottom. The northern hemisphere of Phobos is visible from about 30° above the equator (Phobos' orbit plane), with the side of Phobos facing Mars at the lower right. Phobos presents an illuminated area of about 17 km from top to bottom and 23 km across. The rim of Stickney, the largest crater on Phobos, is seen at the lower left, with a large network of grooves radiating from it. A large, 2-km diameter crater with a slumped wall is seen just below the middle of the picture. [343A27, 29, 31 (P-19133)]

High Resolution View of Grooves on Phobos. This picture shows a northern area on Phobos which is dominated by grooves. An area near the terminator (7.5 × 9.0 km) is seen with visible features as small as 20 meters. Craters of all sizes abound, with a significant portion formed later than the grooves. The grooves radiate from the antipodal point of Stickney, and are probably surface fractures caused by the impact that formed this large crater. Possible outgassing of volatiles during formation could have caused the raised rims along the fractures by ejecting regolith. [246A06]

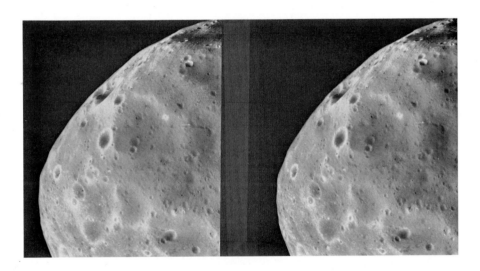

Stereoscopic Views of Phobos. The upper pair shows the side facing away from Mars at a range of 500 km from the orbiter. The large craters near the limb are about 4 km across and a few hundred meters deep. The lower pair shows the side facing Mars at a range of 300 km. The grooves are radiating from Stickney and are tens of kilometers long, hundreds of meters wide, and can be tens of meters deep. [Upper left 246A76, Upper right 246A66, Lower left 343A08, Lower right 343A25]

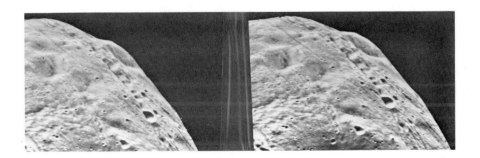

Phobos Overflying Ascraeus Mons. This spectacular picture, taken by Viking Orbiter 2, is the first picture ever taken showing such detail on both a satellite and primary planet. Viking Orbiter 2 was about 13 000 km above the surface of Mars and about 8000 km above Phobos, which increases the apparent size of Phobos relative to features on Mars. Phobos is about 22 km across, and Ascraeus Mons is over 300 km across at its base. The complete outline of Phobos is seen from direct and reflected sunlight. Transit pictures such as this are used to determine the size and shape of the satellite as well as improve the map coordinates of features on Mars registered near the satellite's image. A unique tie between Mars surface (map) coordinates and inertial space can be made when the inertial positions of the satellite and spacecraft are known accurately. [304B88]

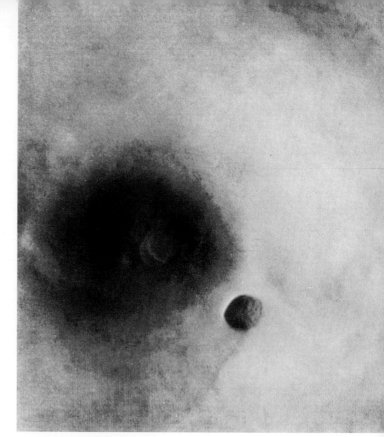

Phobos Overflying the Mouth of Ares Vallis. These mosaics of pictures from Viking Orbiter 1 show Phobos passing beneath the spacecraft with the surface of Mars in the background. These mosaics, taken about a minute apart, show an apparent motion of Phobos across the surface of Mars of about 50 km. Orbiter 1 was 13 700 km above the Margaritifier Sinus region of Mars, and Phobos was 6700 km from the spacecraft. Phobos, four times darker than Mars, appears black against Mars in these unenhanced pictures. This region of Mars contains chaotic terrain along the equator; it is near the head of Ares Vallis, a major channel leading to Chryse basin where Viking Lander 1 is located. [451A03-10; 1° N, 19° W]

Phobos Shadow Transit over the Viking Lander 1 Site. The passage of the Phobos shadow over Viking Lander 1 was imaged simultaneously from Viking Orbiter 1 and Viking Lander 1. The time of shadow passage, as observed by the lander, was used to locate the position of the shadow (and therefore the position of the lander) in the orbiter pictures. This picture shows the shadow of Phobos (~60 × 120 km across) in the Chryse Planitia region a few kilometers directly north of Viking Lander 1. To the left of bottom center is Maumee Vallis, approximately 420 km southwest of the lander's location. [463A21]

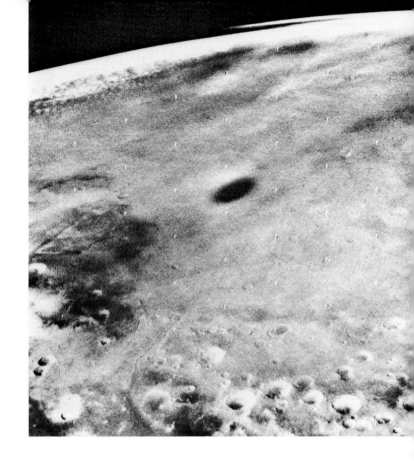

Deimos from Near and Far. A two-picture photomosaic showing the complete side of Deimos visible from Viking Orbiter 2 is on the left, and a high resolution three-picture mosaic of a small area near the terminator is on the right. The two-picture photomosaic, taken at 500 km, shows a smooth surface with limited cratering and a few large flat areas. No linear grooves are seen; however, bright patches of material near the intersection of the large flat areas are visible. The three-picture photomosaic taken at about 50 km gives a completely different view of Deimos than does the two-picture (lower resolution) photomosaic. A surface saturated with craters and strewn with boulders is revealed by the factor-of-10 increase in resolution. Craters have been partially filled or covered by regolith, which gives a smooth appearance to the surface at lower resolution (a range of 500 km or more). A "wind streaking" effect from upper right to lower left probably resulted from a base surge phenomenon when ejecta material was transported and deposited downtrack by the impact of an incoming meteoroid. A few dark-rimmed craters are seen. [Left 428B10-11, Right 423B61-63]

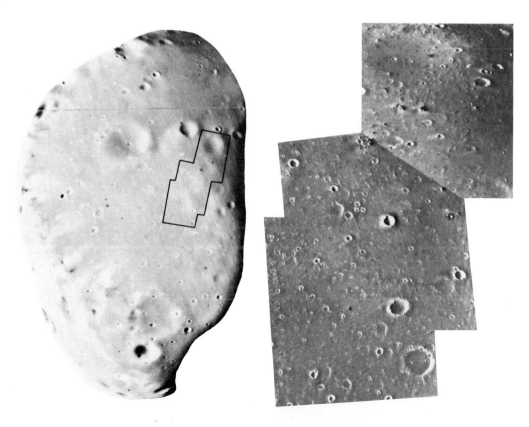

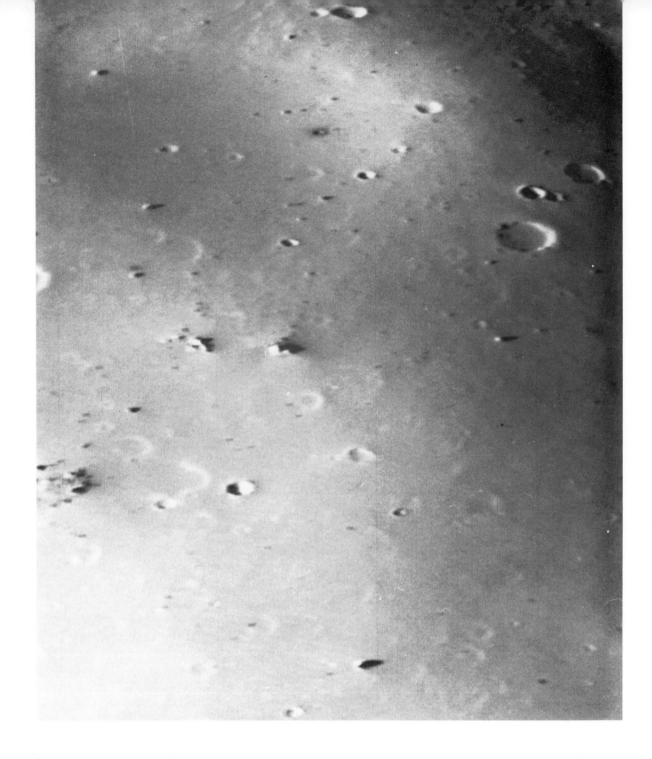

|100 km|

Deimos from 30 Kilometers. Deimos was observed on October 15, 1977 when Viking Orbiter 2 passed within 30 km of the surface. This is one of the highest resolution pictures ever taken of any body in our solar system by an orbiting or flyby spacecraft. The picture covers an area of 1.2 × 1.5 km, and shows features as small as 3 meters. Viking Orbiter 2 would have been visible from the surface of Deimos during this exceptionally close flyby. The surface of Deimos is saturated with craters. A layer of dust appears to cover craters smaller than 50 meters, making Deimos look smoother than Phobos. Boulders as large as houses (10 to 30 meters across) are strewn over the surface—probably blocks ejected from nearby craters. Long shadows are seen cast by these boulders (sunlight is coming from the left). The image was taken when the Sun was only 10° above the horizon. [423B03]

Phobos and Deimos—Similar but Not Identical. In the upper images, the surfaces of Phobos and Deimos are compared at a range of 1400 km. Features as small as 100 meters are detectable. Phobos is viewed at a 90° phase angle and Deimos at a 60° phase angle. Grooves and craters dominate the uniformly dark surface of Phobos at this resolution. Deimos, however, appears to be very smooth, with few craters and with areas of bright albedo. The grooves on Phobos radiate from the large crater Stickney (10 km across) at the left. The bright patches on Deimos are near the intersection of large flat areas. Higher resolution imaging in the bottom images dispels the initial impression of a smooth surface for Deimos, by showing a surface saturated with craters that have been obscured by regolith. [Upper left 357A64, Upper right 413B83, Lower left 244A05, Lower right 423A61]

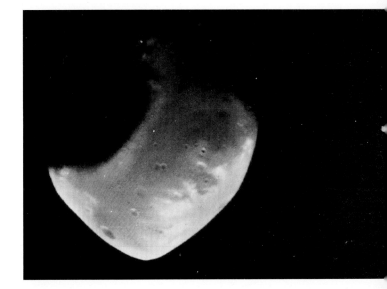
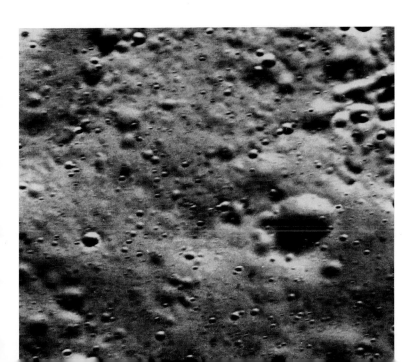

a

Phobos and Deimos in Color. Color pictures of the two Martian moons have confirmed Earth-based spectra by also showing both satellites to be gray. The Viking imaging data showed the surfaces to be uniformly gray over the complete surface to a resolution of a few hundred meters. No significant color differences were seen on either surface, including areas around craters and those within the bright albedo features on Deimos. The color indicates composition is of a carbonaceous chondritic material. Phobos (a) is at a range of 4200 km, and Deimos (b) is at a range of 2100 km. In these pictures, color differences have been exaggerated. Most of the color differences are due to noise or are artifacts of the processing, especially around craters and the limb. [Left 357A03-07, Right 355B01-09]

b

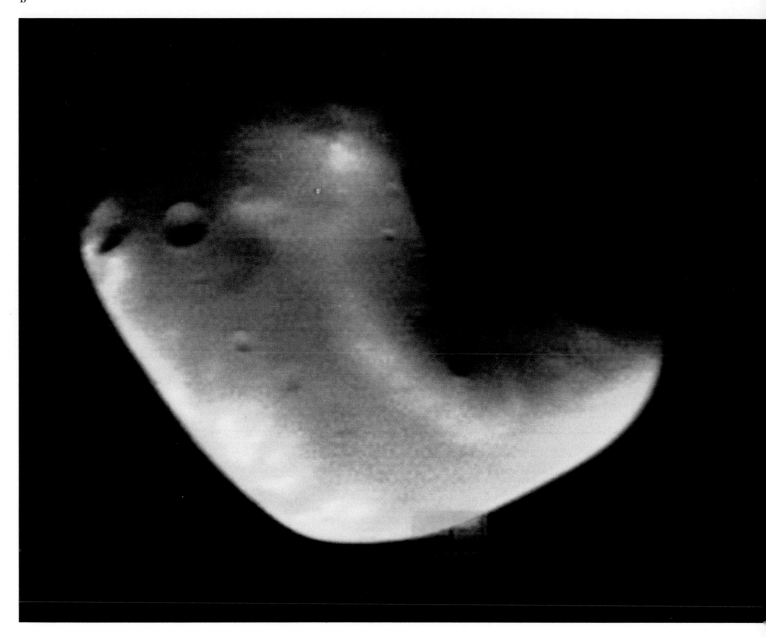

SURFACE PROCESSES

THE MARTIAN SURFACE has been subjected to a wide variety of processes, collectively termed gradation, throughout its geological history. The net effect of gradation is to bring planetary surfaces to a common level by eroding topographically high areas and filling in low areas by deposition. Thus, gradation involves the weathering, erosion, transportation, and deposition of surface materials by wind, water (frozen or liquid), and gravity.

Even before spacecraft were sent to Mars, telescopic observations showed that dust storms are common, and it was speculated that these storms could alter the surface. When Mariner 9 arrived at Mars, a major dust storm had obscured the surface of the planet. After the dust storm cleared, the Mariner cameras revealed a wide variety of landforms related to aeolian (wind-related) processes, including dune fields, yardangs, and shifting albedo patterns consisting of light and dark streaks. The Viking orbiters and landers have provided much additional information on both aeolian processes and landforms.

In the tenuous atmosphere of Mars, much stronger winds than those on Earth are required to pick up particles and set them into motion. Winds of some 150 kph are estimated as minimum for initiation of particle movement. Viking orbiter pictures show several areas in which storms seem to originate; these areas include Daedalia, Hellas, and Syrtis Major, which also display numerous "streaks" associated with craters. Streaks appear to be zones in which fine-grained particles are redistributed in response to wind patterns generated around craters and other landforms.

Some areas on Mars appear to be zones of deposition for windblown particles, as evidenced by enormous dune fields. These areas include the north polar region, the floor of the large impact basin, Hellas, and the floors of other smaller impact craters. The most spectacular of the dune fields, those at the north pole, are discussed in the section *Polar Regions*.

Wind-eroded features include yardangs and grooves etched in some plains. Because the atmosphere is very thin the wind speeds needed to move particles are much higher on Mars than on Earth, so that the grains travel much faster once set into motion. Consequently, when they strike other particles and bedrock surfaces, they have a greater erosion capability than they would have on Earth.

Mass wasting is the downslope movement of materials, primarily caused by gravity, and is seen as landslides, avalanches, and soil creep. Its effectiveness is controlled by factors like cohesion of the material, steepness of slope, gravity, and the presence of lubricants such as liquids and volatiles. Mariner 9 and Viking pictures show many features that can be attributed to mass wasting. Mass wasting along the walls of Valles Marineris has produced some of the most spectacular landslides observed anywhere.

Surface and near-surface processes that occur in the vicinity of former and existing ice regions are referred to as periglacial processes. Although periglacial features and related phenomena have not been positively identified on Mars, it is reasonable to expect them in view of the low temperatures and the probable existence of subsurface ice in some regions. "Etch" pits, polygonal ground, and rock "glaciers" are among the features observed from orbit that may be related to periglacial processes on Mars.

Sand Dunes and Landslides in Valles Marineris. A 40-km-long field of sand dunes (dark area in lower left) and a massive avalanche (middle of mosaic) are seen here on the floor of Gangis Chasma, one of the branches of Valles Marineris. In this region, the walls have been modified by landslides. Debris flows are numerous, as are jumbled masses of debris below the cliffs. Wind may be an effective agent in removing debris that has slumped into the canyon. The canyon thus enlarges itself by the combined processes of slumping and wind excavation. [P16941; 7° S, 45° W]

Details of Valles Marineris Sand Dunes. An enlargement of the dune field in the preceding picture is presented here to show individual dunes about 500 meters across. The wind appears to have been blowing from the west and leading dunes to the east appear to climb the canyon wall. [P16950; 7° S, 45° W]

Landslide in Noctis Labyrinthus. This landslide mass completely fills the floor of the canyon. The canyons in this area appear to be graben that resulted from crustal extension with subsequent widening and modification by landslides. [46A19-22; 10° S, 96° W]

Small Dune Field in Kaiser Crater. Craters and other topographic depressions are natural traps for windblown sediments. The crater shown here is typical of many that have been photographed from orbit. [94A42; 46° S, 339° W]

N

20 km

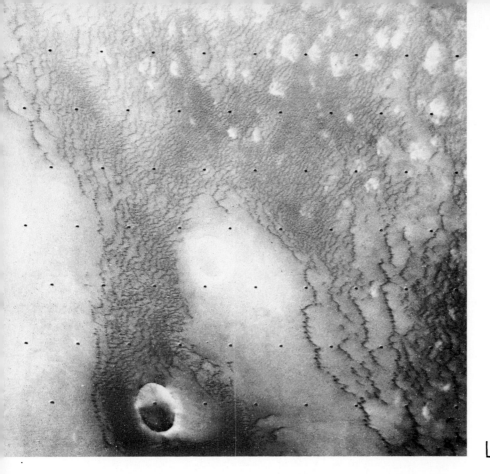

Part of the Dune Field in the North Polar Region. The dune field covers an area of at least 3500 km² and is composed of barchan (crescent-shaped) dunes. In the area shown here, the dunes are aligned in ridges that appear to be transverse to the prevailing wind. From the relation of the dune field to the crater at the bottom of the picture, the prevailing winds seem to be from the west (left side of picture). [59B65; 76° N, 88° W]

Barchan Dunes at Edge of North Polar Cap. This figure shows the well-defined lines of individual barchan dunes. The wind direction is from left to right. [58B22; 75° N, 53° W]

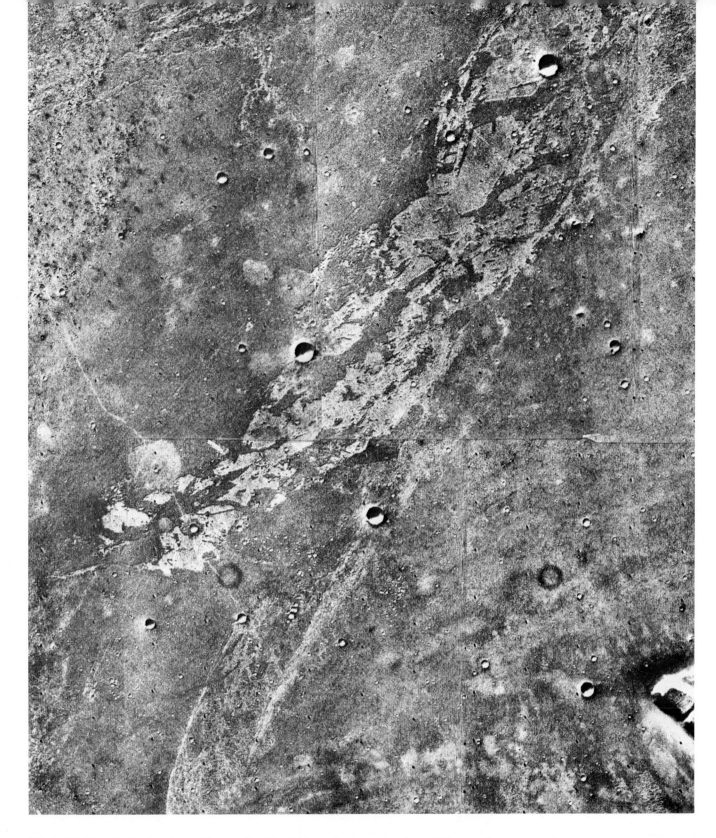

"Etched" Terrain in Southern Chryse Planitia. This etched terrain shows light-toned, angular depressions in southern Chryse Planitia in the area where Tiu Vallis empties into the Chryse basin. The etching process that removed the dark plains material may be the result of cavitation or plucking during active channel formation or wind deflation. Many small, volcano-like features occur in this region. The arrow points to one of these features, a low mound with a summit crater. This feature (also discussed in the *Volcanoes* section) lies on a sinuous line of unknown origin; the line may be the trace of a fracture or possibly a dike. [211-4990; 19° N, 35° W]

Northern Contact of Chryse Planitia. Chryse Planitia "plateau," the mottled light surface at the bottom, is shown at its contact with the darker plains. Irregular pits on the plateau (lower right) suggest formation by collapse; the scalloped scarp of the plateau seems to result from scarp retreat and the connection of the irregular pits. The morphology of the pits and scarp resembles thermokarst features on Earth that result from the melting of ground ice and the subsequent settling of the ground. [211-4994; 23° N, 36° W]

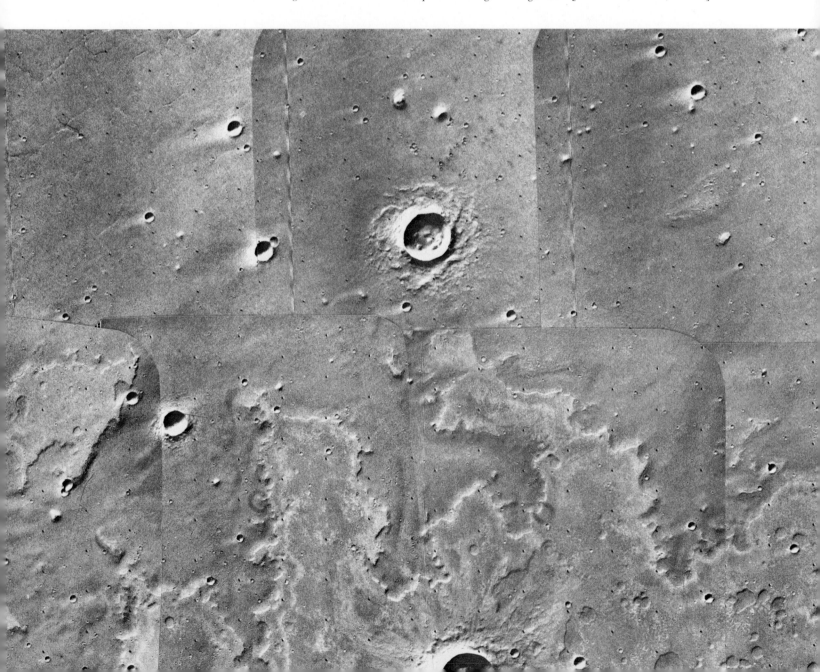

Concentric Flow Features at the Foot of Olympus Mons Scarp. These flow features are more like those typically developed on avalanches and landslides. The unit on which they occur is probably material formed by landsliding on the scarp front. This process may have played a major part in developing the scarp around the volcano. [48B04; 23° N, 138° W]

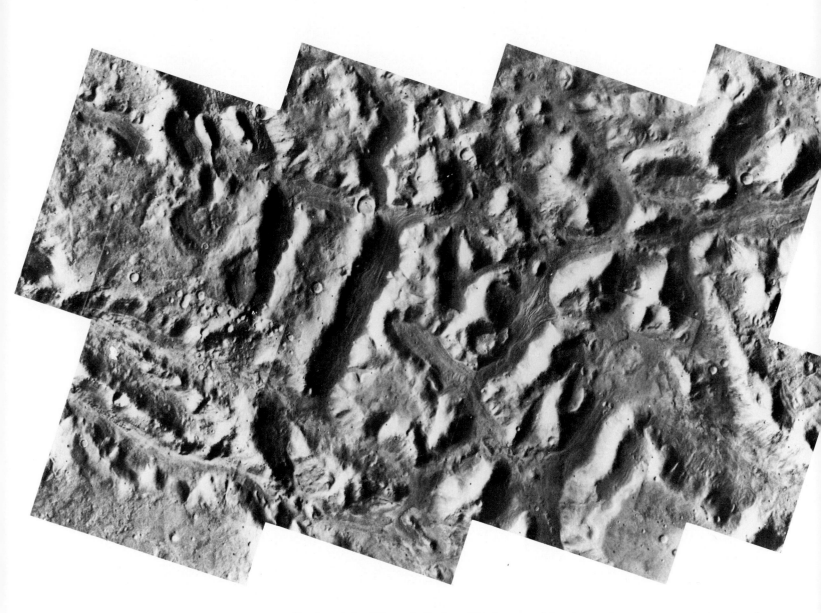

 N

| 10 km |

Mosaic of the Nilosyrtis Region. This is a transitional zone between an ancient cratered terrain to the south (bottom) and sparsely cratered terrain to the north. In many of the low-lying areas there are sub-parallel ridges and grooves that suggest creep of near-surface materials. They resemble terrestrial features where near-surface materials flow *en masse* very slowly, aided by the freeze and thaw of interstitial ice—water frozen between layers of ground materials. This is additional evidence suggesting the presence of ground ice in the near-surface materials of Mars. [P-18086; 34° N, 290° W]

Flow Structures in Ancient Cratered Terrain East of Hellas. Mass-wasting structures around positive features extend up to 20 km from the source. The aprons are not composed of discrete lobate flows, as would be expected if they were formed by landslides, nor are they talus deposits close to the angle of repose; surface slopes are probably less than 10°. Instead, these features may be the result of slow creep of debris containing interstitial ice. [97A62; 41° S, 257° W]

Chaotic Terrain North of Elysium. The plains of the south (lower half of this mosaic) appear to have partly collapsed and then eroded so that only isolated remnants remain. Collapse may have occurred as a result of removal of subsurface ice. A process of planation appears to have removed materials down to a specific depth and created a new planar surface at that depth. It is unclear what the erosive mechanism was or where the material went. [211-5274; 33° N, 213° W]

Contrasting Terrain West of Deuteronilus Mensae. (a) The smooth areas shown may be either debris mantles or remnants of older terrain. In the textured areas, the linear markings may mark the position of former escarpments—the outline of smooth areas. [52A31-44; 44° N, 352° W]

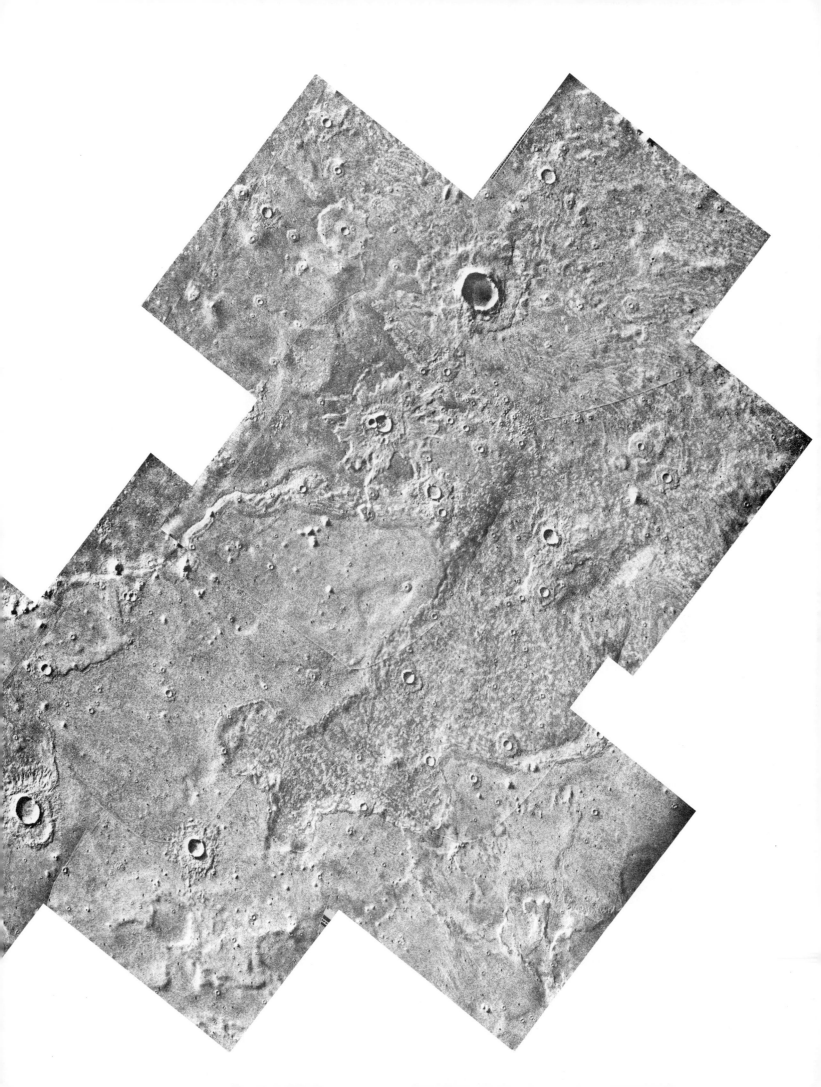

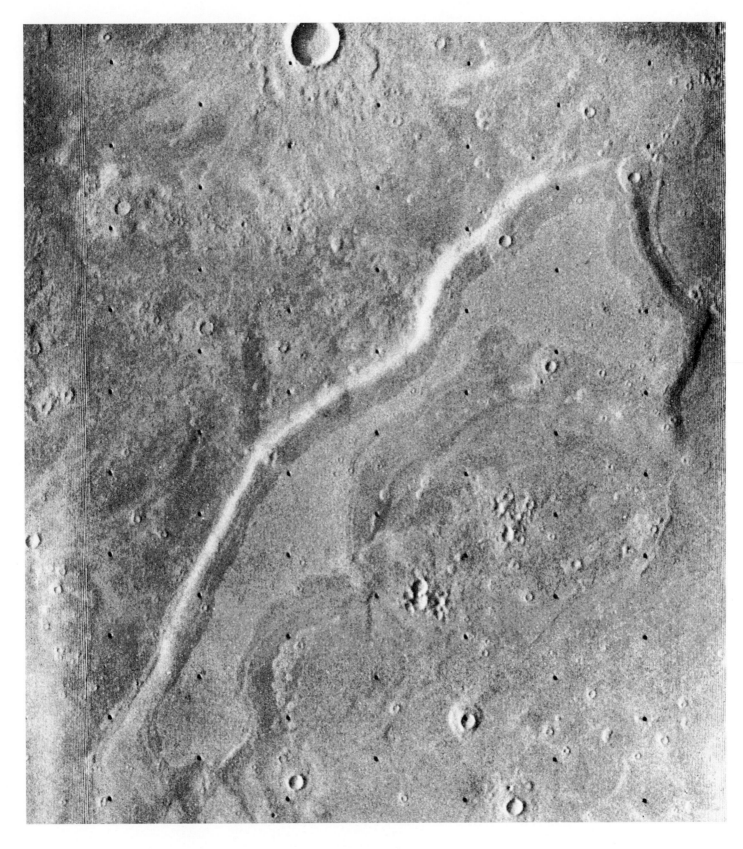

Contrasting Terrain West of Deuteronilus Mensae. (b) A view is shown of part of the Cydonia region of Mars, a 65-km-long remnant of the same plateau unit shown in (a). [26A72; 45° N, 7° W]

N

|— 5 km —|

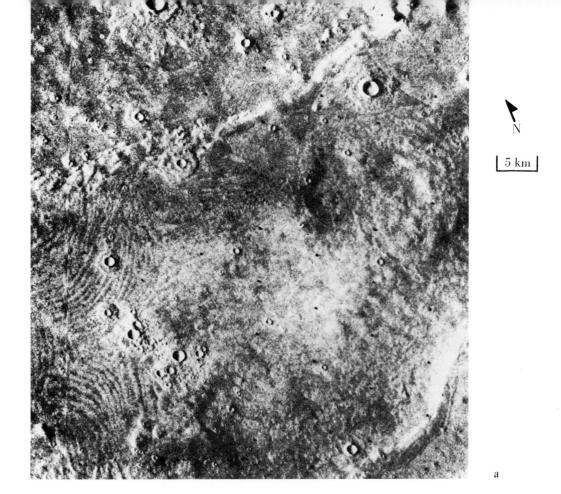

Striped Ground. (a) Geometric markings resembling contour plowing in the Cydonia region are seen, and consist of low ridges and valleys about 1 km from crest to crest. The features may mark successive positions of the retreat of an escarpment during removal of a plateau or mantling unit. (b) In this high resolution image of striped ground similar to that in (a), the parallel markings are caused by low ridges and, less commonly, shallow depressions. [(a) P-17599; 46° N, 350° W, (b) 11B01; 50° N, 289° W]

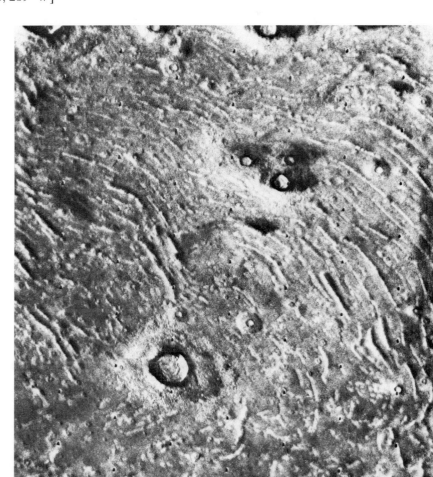

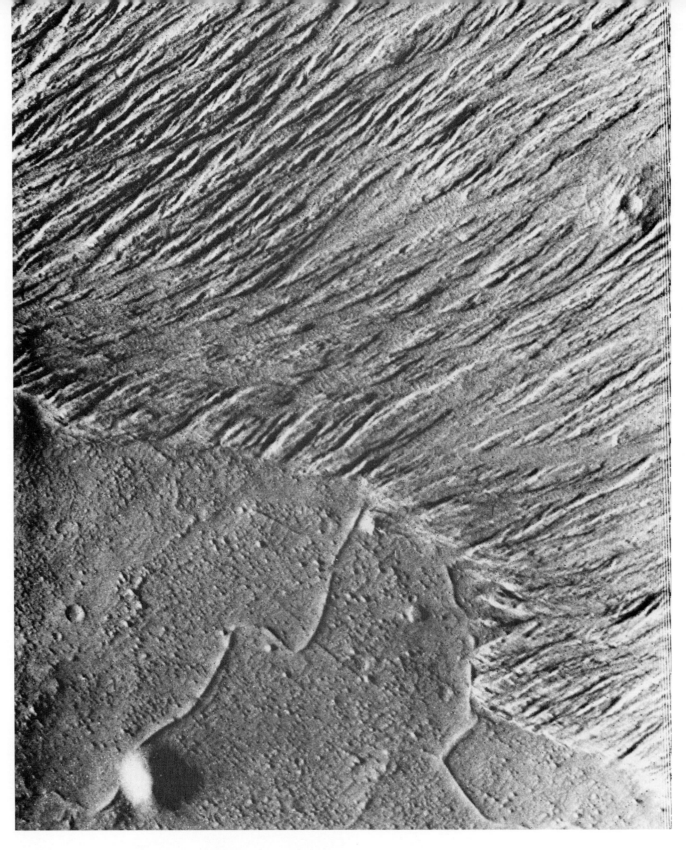

1 km

Highly Textured Eroded Surface. The upper half of this image shows a layer of relatively erodable material that is being sculpted and swept away by the wind. In the lower left a more resistant older surface has been exposed which is dominated by small hills and sinuous, narrow ridges. The hill at the bottom may be of volcanic origin. The narrow ridges are especially puzzling. It has been suggested that they may be dikes but their extensive continuity and ridge-like surface forms argue against this. An alternative, but weaker, hypothesis is that they may be eskers. [724A22; 2° S, 210° W]

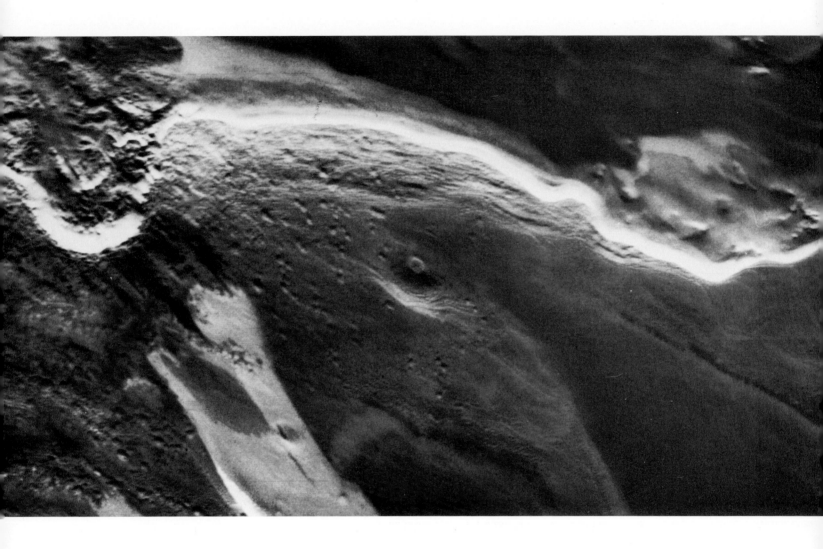

POLAR REGIONS

THE APPEARANCE of the polar regions contrasts sharply with the rest of the planet, partly because of varying amounts of frost cover and partly because of some highly distinctive terrain not found elsewhere. Both poles have a cap of frozen carbon dioxide that advances and recedes with the seasons. In the north a small permanent residual cap left in midsummer is composed of water ice. The composition of the small residual cap left at the south pole is not known. The residual northern cap is substantially larger than the residual southern cap, so much so that the unique polar terrains of the north are rarely seen without some frost cover. The polar scenes are all from Viking Orbiter 2, which was placed in a high-inclination orbit specifically to view the poles. Because its periapsis was in the high northern latitudes, the highest resolution photographs are of the north.

The most distinctive geologic features of the polar regions are thick, layered deposits that cover much of the surface poleward from 80°. The layering is best seen where the frost has been preferentially removed such as on terraces and on walls of valleys within the deposits. The layers, which range in thickness from several tens of meters down to the resolution limit of the available photography, can be traced laterally for considerable distances. Unconformities occur but are relatively rare. In the north, the layers rest on sparsely cratered plains; in the south they rest on old cratered terrain. The layered terrain is almost completely devoid of impact craters. Either resurfacing by erosion or deposition is at a rate that is high compared with the impact rate, or the impact craters "heal" relatively quickly by flow or infilling.

The layered deposits are believed to be accumulations of volatiles and wind-blown debris, with the layering caused by variations in the proportion and absolute amounts of these two components. If this interpretation is true, then the layered deposits preserve a partial record of the history of atmospheric activity, and hence climate, in the recent geologic past.

A vast belt of dunes, several hundred kilometers across, surrounds the layered terrain in the north. In some areas, the dunes form a nearly continuous sheet that almost completely masks the underlying topography. In other areas, particularly around large topographic features, the sheet is discontinuous and breaks up into strings of crescentic dunes or isolated forms. Dune fields of comparable continuity do not occur around the south pole, although numerous dark splotches on the surface in the high southern latitudes are probably local dune fields.

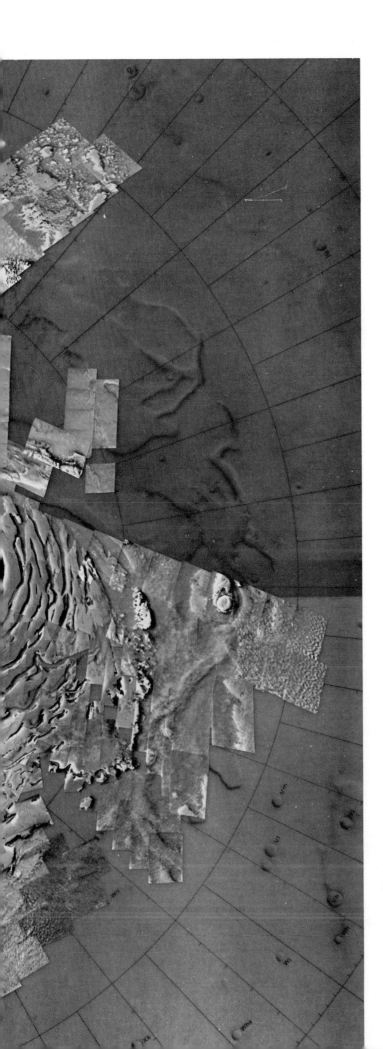

| 100 km |

Photomosaic of North Pole. This photomosaic consists of some 300 Viking Orbiter 2 frames. Around the north pole, curving in huge arcs, are dark bands where polar frosts are absent. A giant ring of sand dunes surrounds the polar region between 80° and 70° N latitude. [211-5359]

Remnant North Polar Cap Detail. This high-resolution, closeup view was made by combining three black-and-white images obtained through color filters. Above center in the picture is a giant cliff about 500 meters high. Layers averaging 50 meters in thickness are seen in the cliff face and surrounding areas, which are highlighted by occasional white patches of frost. The regularity of the layering suggests that it comes from periodic changes in the orbit of Mars—a relationship that, on Earth, may be at least partially responsible for ice ages. These orbit changes may affect the frequency and intensity of global dust storms, in turn varying the amount of material available to form layered terrain. The cliff is apparently an erosional feature; the variety of scarps shows the complexity of erosion in the polar regions. Dune-like features (dark areas with a rippled texture), possibly formed from material eroded from the layered terrain, can be seen at the center and at the right of the picture. Just above the scarp, the polar ice layer is very thin and patchy; in other places it appears to be considerably thicker. The maximum thickness of the polar cap has not been determined. [75B52, 75B56, 75B58 (P-18459); 84° N, 237° W]

Layered Deposits Partially Covered by Frost near the North Pole. The light and dark pattern is caused largely by the presence or absence of frost. The layers are best exposed on southward facing slopes, which are generally without frost and hence are dark. Although sequences of layers can commonly be traced unbroken for considerable distances, breaks in the sequence do occur. This pair of pictures includes an example of an angular unconformity where one set of deposits truncates another at an angle. In (a), the unconformity is in the upper right center. In (b), an enlargement of the area with the unconformity, the fine scale layering of the sequence shows more clearly. [56B84; 80° N, 339° W]

A Dune Field in Borealis Chasma. Dark dune-forming materials appear to have been transported away from the pole in a curving stream extending from the top of this frame. They are accumulated in an approximately triangular dune mass that occupies the center of the mosaic. The sinuous ridges in the dune mass rotate in a clockwise direction through an angle of approximately 45° from the northern to the southern margin. The discontinuous dark texture on the right side arises from partial dune cover. Perennial ice is visible near the top of the frame and associated with the crater near the bottom of the frame. The bright patch near center right may be a cloud. [58B21-34; 48° N, 52° W]

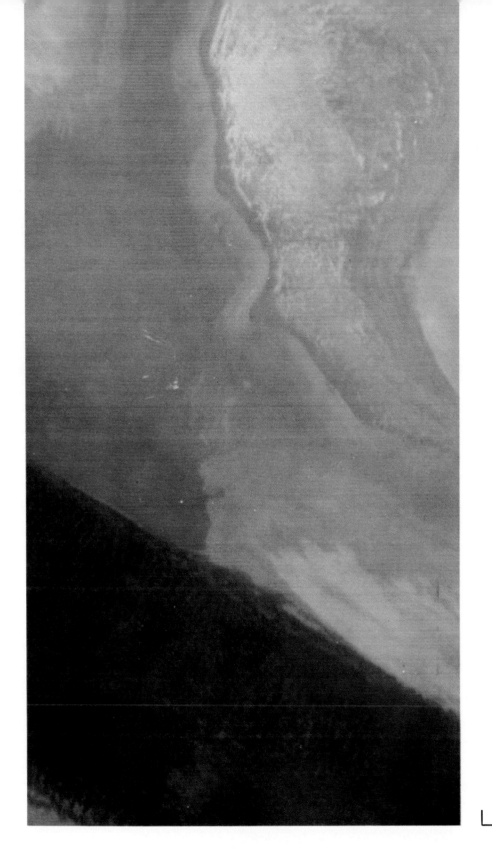

5 km

Sand Dunes at the Rim of the North Polar Cap. The dunes form a sharp-edged, dark band near the bottom of this image. Martian sand is dark, unlike Earth sands which are usually light colored. This shows the minerals in Martian rocks most resistant to erosion are the dark ones. The center of the image shows a flat desert region. At the upper right are a region of mottled terrain of unknown origin, a strip of layered terrain (its layering clearly visible in this view), and a pinkish-white region of polar frost. [IPL, ID:I2398AX; 81° N, 83° W]

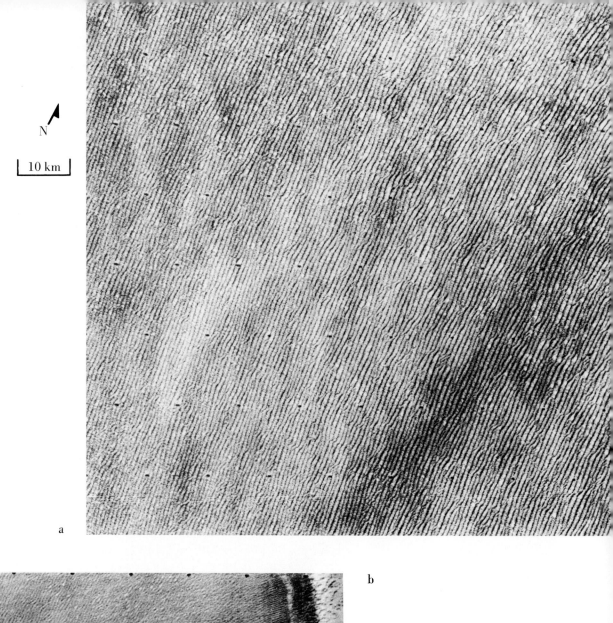

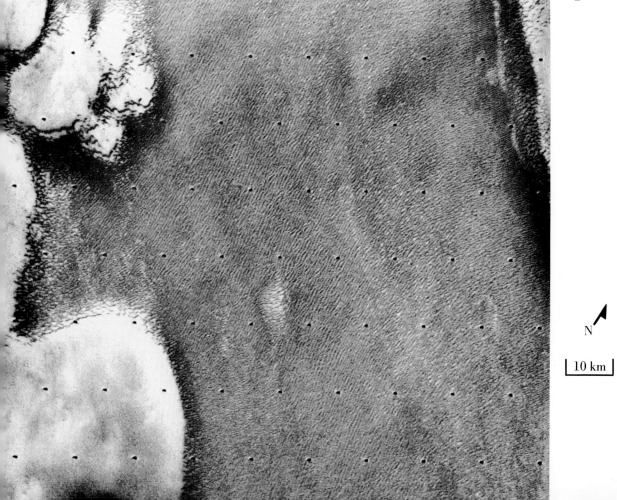

Widespread North Polar Edge Sand Dune Fields. (a) Dunes here have a consistent trend (approximately north-south) with minor sinuosity, branching and merging. Vague circular forms are probably buried craters, and bright spots within the ridges are ice deposits. (b) Dunes with much more variation in direction also occur; a shorter wavelength and greater sinuosity appear in this dune field which adjoins—and in places appears to be mantled by—frost deposits. Vague circular forms again are probably buried craters. The bright patches of ice near the upper left are associated with a distinct change in the dune pattern, possibly indicating that the deposits of ice preceded the development of the present dune pattern. (c) Transition from a transverse ridge structure to isolated linear and equant dunes. [(a) 59B32; 81° N, 141° W, (b) 58B01; 80° N, 120 ° W, (c) 58B28; 78° N, 50° W]

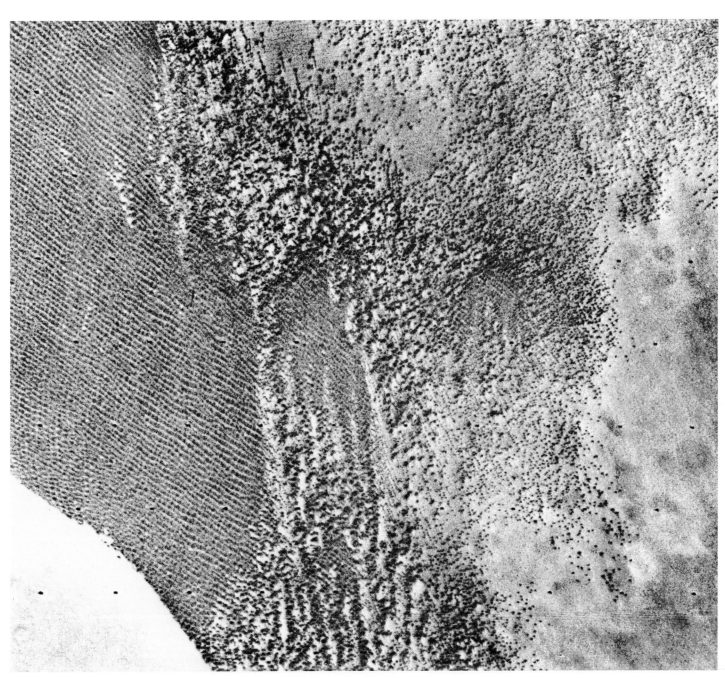

c

Photomosaic of South Pole. At the left of this photomosaic is the remnant south polar cap of Mars. Until recently, evidence suggested that its composition was water ice like the remnant north polar cap. New temperature measurements, however, suggest that it may be carbon dioxide ice. Extending from beneath the polar cap to the bottom of the frame are large, lobate expanses of glacioaeolian deposits with wind-scoured surfaces. At their northern margins, these deposits overlap and partially fill a number of craters. They also mantle the entire southern wall of a huge impact basin, 800 km in diameter, which is approximately at the center of the photomosaic. The unburied part of the basin rim, or rampart, forms a mountainous, semicircular arc, with plains in the interior and a rugged landscape of large craters stretching to the north. At the smallest scales, the polar terrains exhibit mysterious patterns and textures which can possibly be attributed to volcanic and wind action, and to cyclical climate change. [383B04-75, 211-5541]

Layered Materials Resting Unconformably on Cratered Terrain near the South Pole. Layered material with a smooth, uncratered surface partly covers a 40-km diameter crater in the upper half of the picture. Strings of secondary craters around the larger crater are also transected by the layered deposits. [383B50; 81° S, 271° W]

20 km

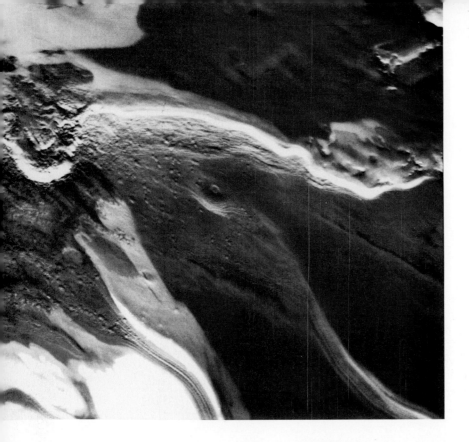

Secondary Craters in Layered Materials Close to the South Pole. Layered deposits are shown in the lower half of the picture but have been eroded away in the upper half to form a low scarp to the north which is illuminated by the Sun. Numerous secondary craters occur in the layered deposits around a partly eroded crater. The relations suggest that the crater formed after the layered deposits but before the erosional episode that formed the north-facing scarp. Part of the remnant cap is visible in the lower left. [421B79; 85° S, 352° W]

Sinuous Ridges on the South Polar Plains. The origin of these ridges is unclear. They branch and rejoin like river channels, and somewhat resemble terrestrial eskers (ridges formed by deposits from subglacial rivers), but a volcanic or tectonic origin is more likely here. Similar features occur elsewhere on the planet, such as on the floor of Argyre. [421B53; 78° S, 40° W]

Pitted Terrain near the South Pole. Some areas peripheral to the layered deposits at the south pole appear to be deeply etched with numerous irregularly shaped depressions inset into a formerly planar surface. The depressions may form by collapse after melting of ground ice or, alternately, they may be simply deflation hollows formed by removal of material by the wind. Similar features do not occur at the north pole. [390B90; 77° S, 74° W]

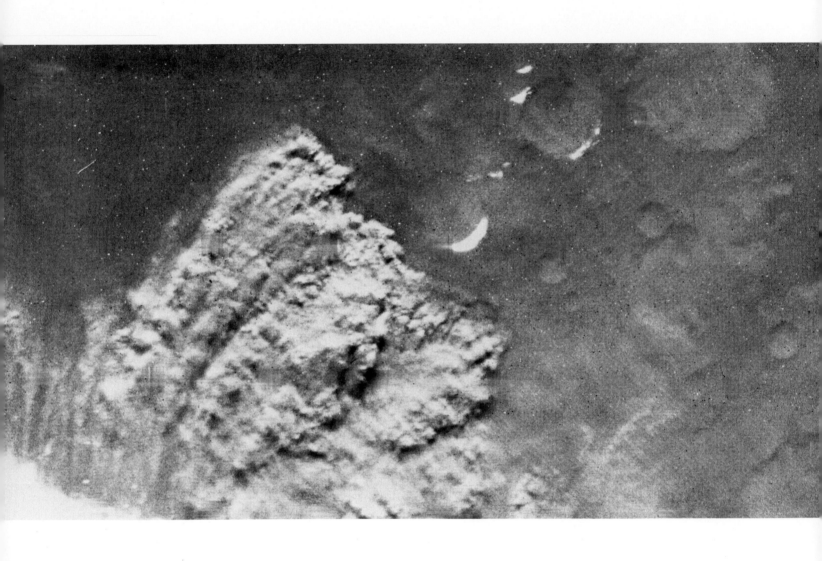

THE ATMOSPHERE

THE MARTIAN ATMOSPHERE is persistently hazy. The haziness is due to the scattering of light by suspended dust and condensate particles. This haze causes the Martian sky to be gray to yellow instead of blue as on Earth; the blueness of Earth's sky is due to the scattering of light by air molecules. Superimposed on the Martian haze are various types of local condensate clouds and fogs. At times, dust storms raise great yellowish clouds that stand out against the haze and ultimately contribute to it.

Because the axis of Mars is tilted with respect to its orbit plane, the Martian atmosphere undergoes seasonal changes analogous to those on Earth. Viking spacecraft arrived just before northern summer solstice. Approach images show a relatively dense haze covering the northern hemisphere and a much clearer atmosphere in the south. With the beginning of southern spring, an even denser haze blanket formed over the southern hemisphere, largely obscuring the surface even from vertical view. Later this southern haze thinned but, as southern summer approached, dust storms again obscured large areas.

Northern latitudes were obscured by condensate clouds and hazes during fall and winter in that hemisphere. North of about 60° latitude, this "polar hood" was diffuse and featureless and, because of the very low atmospheric temperatures in these regions, is believed to be at least partly carbon dioxide ice particles. The zone between 40° and 60° N was swept by fronts that moved south out of the polar regions; cloudiness was associated with these weather systems.

Images of the Martian limb regularly show a high, layered haze structure extending to more than 35 km above the surface, with individual layers typically extending over large areas. The vertical distribution of light-scattering particles is not directly proportional to the brightness profile in the limb image. This condition is because lower layers are seen along paths of varying length through upper layers. The true distribution of scatterers was calculated, and results revealed the existence of clear layers between the cloudy ones.

The diffuse haze blanket itself is not without structure. In some regions its features include broad longitudinal streaks, cellular lumpiness, and wave trains. Cells, which range in size from about 1 to 10 km, indicate convection within the haze blanket. Wave trains up to several hundred kilometers long are visible in a large percentage of high-altitude frames near the morning terminator. These waves are visible because of the alternate condensation and evaporation of ice crystals in the troughs and crests of a pressure wave traveling through an atmosphere of high static stability.

One prominent type of condensate cloud on Mars forms around the giant volcanic mountains of Tharsis and Olympus. These clouds, evidently formed by orographic uplift, form in late morning and obscure the flanks of the volcanoes up to an elevation of about 20 km, leaving the summits unobscured. In Earth-based observations, these clouds have been known for decades as the "W clouds" because of their repeating configuration. Other types of condensate clouds occur over less than 1 percent of the Martian surface at any particular time. These include convective-like formations, cirrus-like wisps, and low-lying canyon clouds.

Observers using telescopes have known for many years that global-scale dust storms are common when Mars is closest to the Sun in its relatively elliptical orbit. Such a storm enveloped the planet when Mariner 9 arrived at Mars. Two smaller global dust storms were observed by Viking orbiters during the extended mission. The first occurred early in the southern spring, and the other shortly after southern summer solstice. Both storms probably started in the Thaumasia-Solis Planum region, and rapidly engulfed most of the planet. They greatly affected meteorology at the landing sites, and each prevented the acquisition of clear images of the Martian surface for 2-3 months. Several dozen localized dust storms were also observed by the Viking spacecraft. Most of these occurred near the retreating south polar cap or in the region to the south of the canyons on the southeastern slopes of Tharsis.

▷

Water-Ice Cloud on Flanks of Ascraeus Mons. This southern view of the dawn side of Mars was taken during August 1976 by Viking Orbiter 2 as it approached the planet. Because it was winter in the southern hemisphere at that time, the south pole is in the dark. Part of the adjacent seasonal frost cap is visible at the bottom center. The great equatorial canyon system, Valles Marineris, is faintly visible at center right; but hazy atmosphere obscures surface features north of that except for the protruding summit of the giant volcano, Ascraeus Mons. The white feature on its western flank is thought to be a type of water-ice cloud frequently observed in that region. [P-19009]

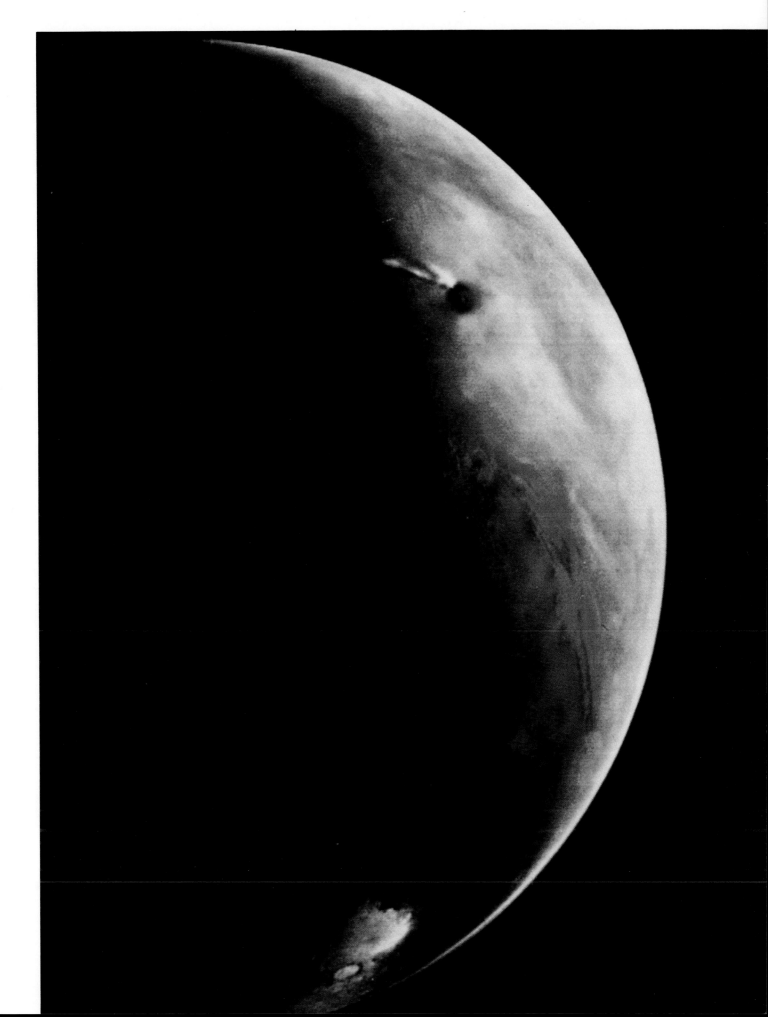

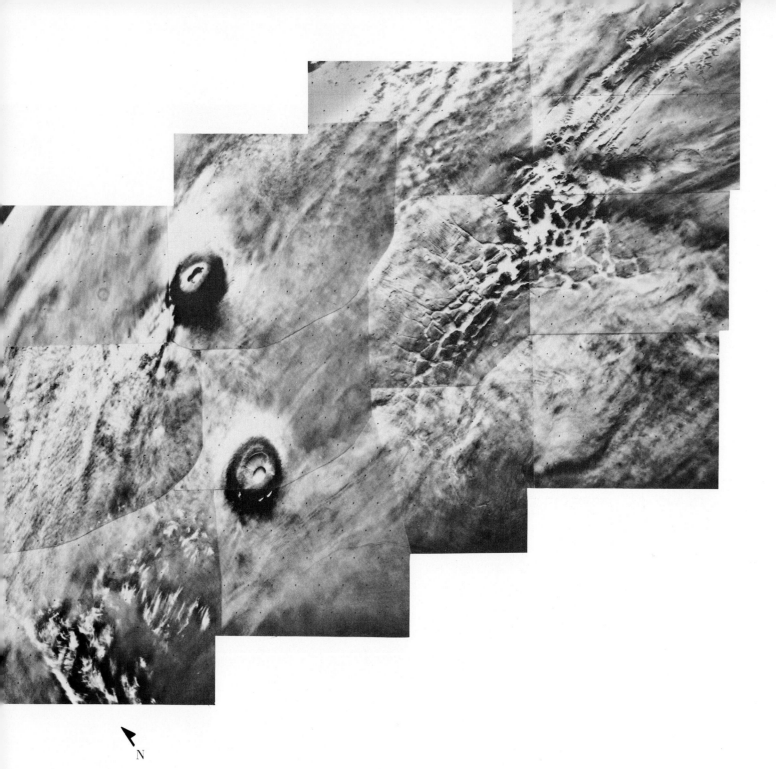

Early Morning Clouds in the Tharsis Montes and Valles Marineris Region. Ascraeus Mons and Pavonis Mons are prominently displayed in this mosaic, and dense cloud blankets cling to their northern slopes. High cirrus clouds lie to the west of Tharsis, and waves are visible in the clouds surrounding the peaks. Bands of clouds appearing to have a cellular structure extend north from the canyon, and the areas within and immediately surrounding the chasm exhibit water-ice fogs. [211-5049; 5° S, 105° W]

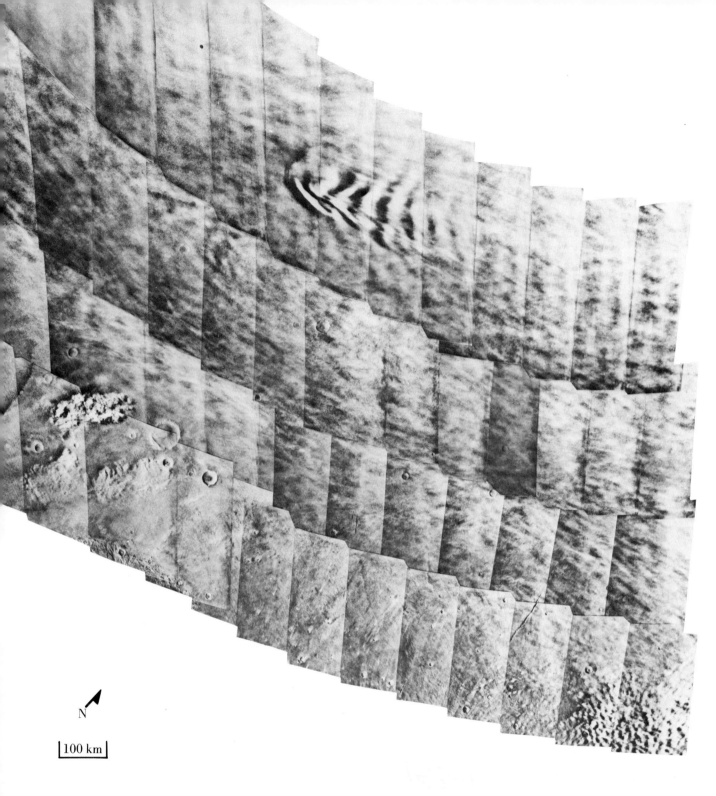

Wave and Dust Clouds in Arcadia Planitia. This mosaic of Viking Orbiter 2 frames shows an area north of Olympus Mons. Surface detail north of 45° is obscured by the polar hood. Well-developed wave clouds, seen at the upper right, are produced by strong westerly winds perturbed by the large crater, Milankovic (55° N, 147° W). The wavelength (distance between crests) of these clouds is about 60 km; their persistence through more than 500 km implies stability in the atmosphere which prevents the dissipation of the waves by turbulence. The dust clouds at the lower left are probably associated with passage of a cold front moving out of the polar hood region. [211-5378; 43° N, 124° W]

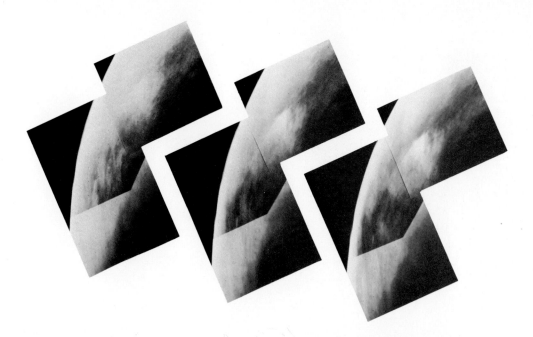

△
Condensate Clouds over the Viking Lander 1 Site. During the summer, the northern hemisphere of Mars is generally quite hazy—as shown in the Orbiter views taken in red, green, and violet light (left to right) from a distance of 32 000 km. Because all colors show some obscuration, the haziness is probably caused by both dust and condensates. The large diffuse cloud near the top center, however, is brighter in violet light than in red, suggesting that it is largely composed of condensates. It appeared over the Viking Lander 1 site in the Chryse basin just a few days before landing. [211-5143; 25° N, 45° W]

Changes in Atmospheric Clarity. These two views in violet light illustrate the dramatic change in the clarity of the atmosphere in the region east and northeast of the Argyre basin during winter in the southern hemisphere. (a) Most of the snow-covered Argyre basin is shown. This was taken just after the winter solstice when solar heating was minimal. (b) This view was taken in late winter when the area had started to warm. The cold southern regions may trap water vapor from the much warmer northern hemisphere to form these clouds, or water vapor may be released from the seasonal polar cap as it retreats. [(a) 34A13, (b) 81B21; 47° S, 22° W]

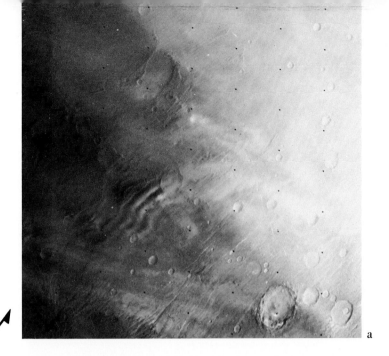

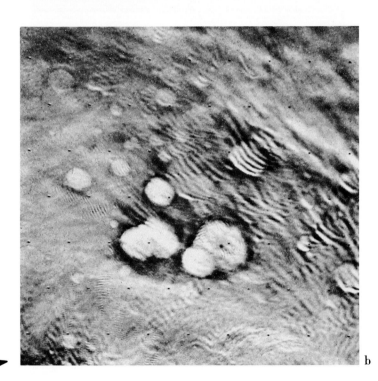

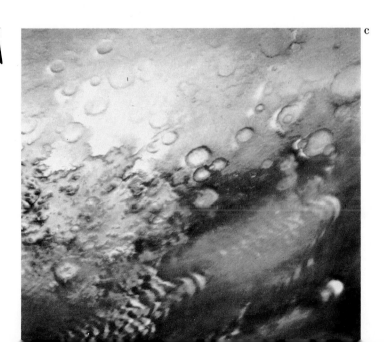

Wave Clouds. Wave clouds form in a stable stratified atmosphere when winds pass over topographic features such as crater rims. The distance between crests (wavelengths) depends on the dimensions of the perturbing feature and on the speed and vertical profile of the wind. (a) These 20-km-wavelength clouds seem to be formed by westerly winds perturbed by the small ridge to the west of the clouds. (b) This complex pattern of waves has wavelengths between 2 km and 15 km, and may be connected with the south polar crater field seen through the haze or perhaps with instability induced by wind shear. The air is quite dusty in the picture, which was taken in red light soon after the onset of the second global dust storm. (c) This view shows waves to the west of Argyre which are associated with a weather system which also produced the Argyre dust storm. [(a) 40A21; 30° S, 88° W, (b) 207B43; 60° S, 154° W, (c) 131B64; 55° S, 65° W]

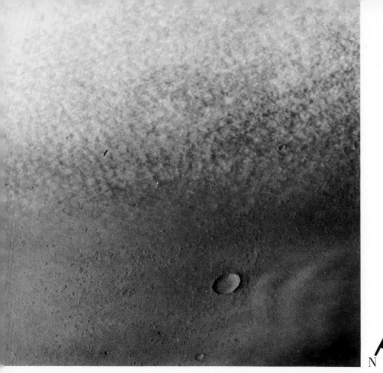 

a b

Cirro-Cumulus and Strato-Cumulus Clouds. Clouds with cellular structure resembling terrestrial cirro-cumulus and strato-cumulus clouds are quite common on Mars, especially in the polar-hood region. Small convective cells, created when the base of the cloud layer is heated by ground radiation, are responsible for the structure. (a) Cellular cloud layers are seen at the edge of the polar hood, viewed from a distance of 15 000 km. Note the lee waves produced by the crater. (b) View, taken from a distance of 1400 km, of cellular clouds in the north polar hood, showing the alignment of the cells into "streets." These features can be produced by vertical wind shear. [(a) 470A07; 40° N, 210° W, (b) 138B53; 73° N, 318° W]

▷

Limb Pictures. Limb pictures (those that include the edge of the planet's disk) show that condensates, and perhaps dust, exist in layers in the atmosphere up to 40 km above the planet's surface. The limb structure in the southern hemisphere is shown in (a) during the early winter and in (b) during the late winter. View (c) depicts the north polar limb and (d) the south polar limb. Both polar views were obtained during the late summer for each hemisphere. [(a) 53A65; 40° S, 40° W, (b) 79B06; 48° S, 253° W, (c) 78B71; 80° N, 346° W, (d) 393B01; 78° S, 84° W]

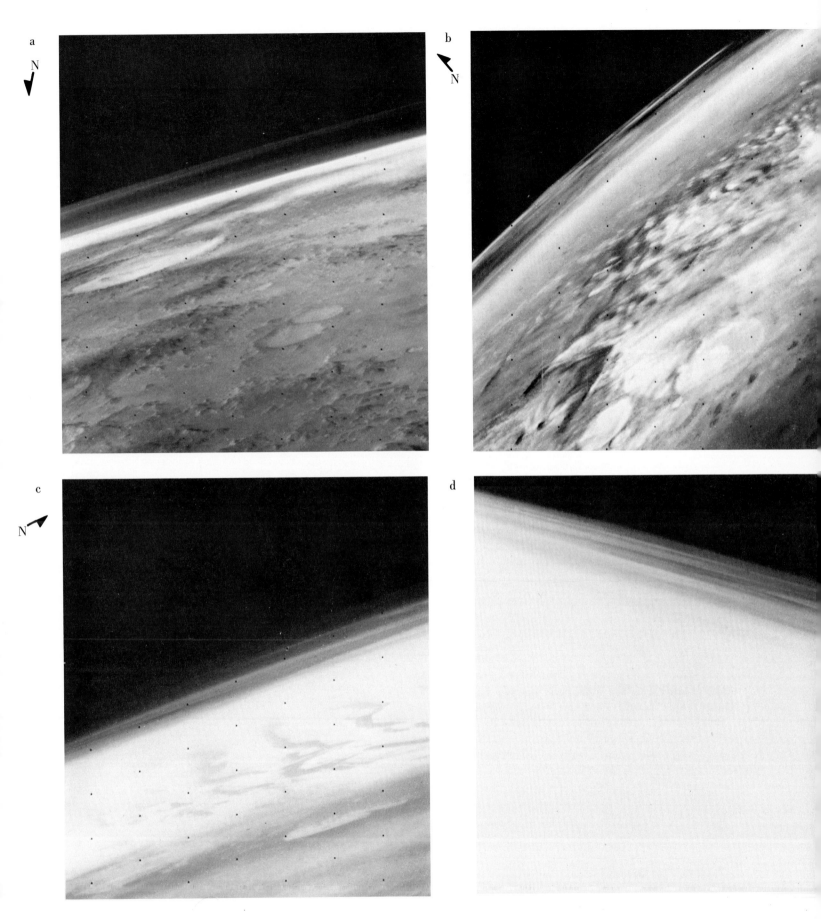

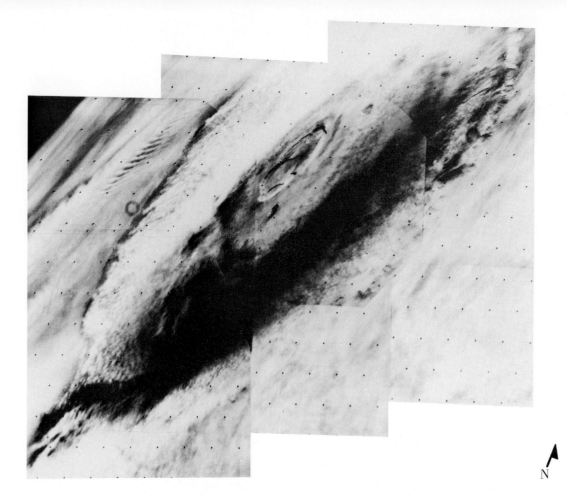

Clouds Surrounding Olympus Mons. In this mosaic, Olympus Mons, wreathed in clouds at midmorning, was viewed obliquely (at an angle of 70° from vertical) from a range of 8000 km through a violet filter. The season is early summer when Olympus Mons receives close to its maximum solar flux. The top of the cloud blanket is about 19 km above the mean ground level and 8 km below the summit. Water-ice, which condenses as upslope air currents cool, is thought to form these clouds. Parts of the cloud cover have a cellular appearance, indicating convection within the clouds. A well-developed wave cloud several hundred kilometers long is visible toward the limb. [P-17444; 18° N, 133° W]

Clouds around Pavonis Mons. Early morning views, taken 3 weeks apart, show Pavonis Mons, the central volcano of the Tharsis Mons receives close to its maximum solar flux. The top of the cloud blanket is about 19 km above the mean ground level and 8 km below the summit. Water-ice, which condenses as upslope air currents cool, is thought to form these clouds. Parts of the cloud cover have a cellular appearance, indicating convection within the clouds. A well-developed wave cloud several hundred kilometers long is visible toward the limb. (a) 40A95, (b) 62A18; 0° N, 113° W]

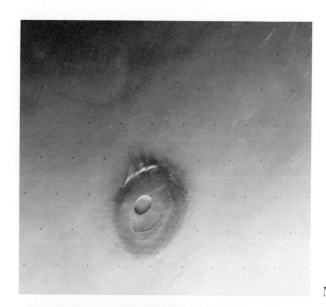

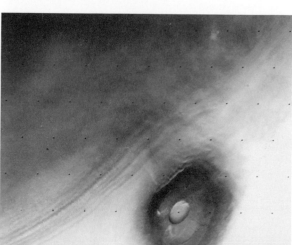

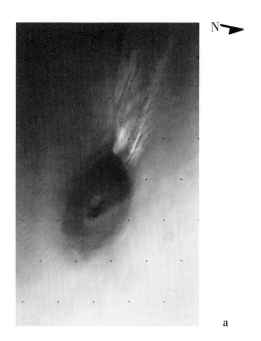

a

Discrete Clouds on Volcano Slopes. Discrete clouds are frequently seen above the slopes of the large volcanoes. (a) The unusual plume cloud was repeatedly seen over Ascraeus Mons in the early morning during the summer. (b) The cloud shown is located over the northwest slopes of Ascraeus Mons; the picture was taken when the local season was early autumn and the time about 2:00 p.m. Picture (c) shows an unusual combination of cirrus-like clouds, thin wave clouds, and a prominent discrete cloud (which may be a turbulent rotor) over Arsia Mons. [(a) 58A12; 11° N, 105° W, (b) 225A05; 12° N, 104° W, (c) 344B88; 9° S, 120° W]

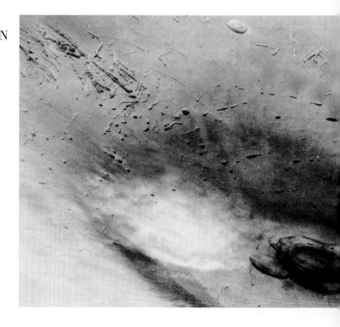

b

c

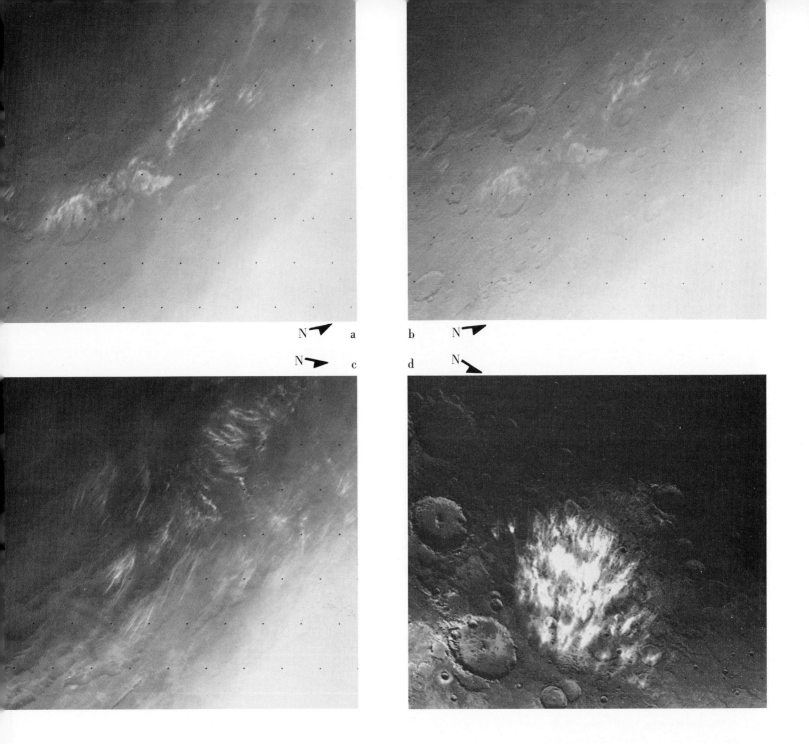

Cirrus Clouds. Clouds resembling terrestrial cirrus clouds are often seen in the Martian atmosphere. That these clouds are condensate phenomena is well illustrated by the greater contrast in (a), taken through a violet filter, than in (b), taken through a red filter at the same time. Without shadows to determine altitudes and a knowledge of temperatures at the proper heights, it is difficult to distinguish water and carbon dioxide ices. It is not improbable that both types of cirrus clouds exist. The group of cirrus clouds in view (c) occurred to the north of the Valles Marineris canyon system; the varying orientations of the clouds may indicate differences in wind direction at the altitudes at which particular clouds occur. Picture (d) shows a bright winter cloud as it appeared over the Electris region. It was observed to recur at the same place on several days during that season. Bjerknes Crater is at the lower left. [(a) 101A10; 6° N, 244° W, (b) 101A07; 6° N, 244° W, (c) 58A02; 6° S, 76° W, (d) 88A03; 42° S, 192° W]

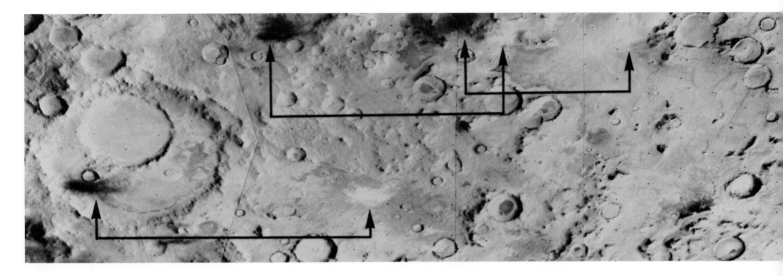

Cloud Shadows. Shadows of clouds may be used to determine the altitudes of clouds. This information, coupled with height profiles of temperature and pressure, can lead to a determination of the composition of a cloud. In mosaic (a), high-altitude clouds are seen over ancient, cratered terrain to the east of the Hellas basin. Arrows connect three small condensate clouds to their shadows, which appear to be about 200 km away. Using simple geometric relationships involving the cloud, its shadow, and the sun elevation angle, one finds the clouds to be at approximately 50 km altitude. View (b) shows a larger (100 km long) cloud south of Valles Marineris, about 50 km above the surface. At this altitude, where temperatures and pressures are low, carbon dioxide is the probable composition. [(a) 97A75, 77, 79, 81; 50° S, 246° W, (b) 318A24; 20° S, 44° W]

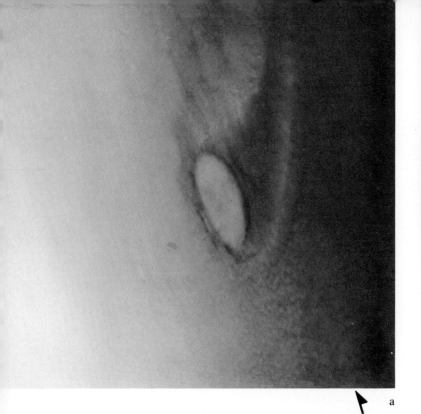

Enigmatic Clouds. These four frames show Martian atmospheric phenomena that do not fit into any of the preceding categories. Picture (a) is an unusual polar hood cloud formation associated with the large crater, Mie. Superposition of lee waves from parts of the terrain around Mie could produce such a formation under appropriate atmospheric conditions. In (b), a cloud in the southern hemisphere can be seen; to catch the Sun's rays the cloud must be high in the atmosphere. Linear, optically thin streaks are seen in the Thaumasia region in (c). Streaky, condensate hazes that have developed near the dawn terminator during the onset of autumn in the southern hemisphere are seen in (d). [(a) 470A05; 48° N, 220° W, (b) 211B60; 55° S, 234° W, (c) 67A06; 39° S, 85° W, (d) 431B03; 26° S, 280° W]

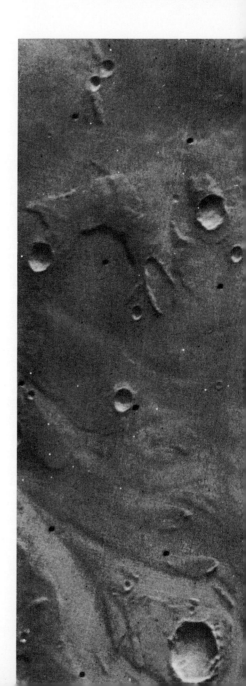

Early Morning Surface Fog. The presence of morning fogs in some crater and channel bottoms is a Viking discovery with possible implications for the future biological exploration of Mars. These early morning views of the Memnonia region were taken one-half hour apart using a violet filter to enhance the contrast of the condensates. The areas marked by arrows are noticeably brighter in the later picture. The fogs indicate specific spots where water is exchanged, probably on a daily cycle, between the surface and the atmosphere. The surface and lower air layers in this region become unusually cold at night because of the thermal properties of the surface. When the surface warms in the morning, it seems that a small amount of water vapor—estimated to be about one-millionth of a meter thick if liquefied—is driven off; this vapor recondenses in the atmosphere, which warms more slowly, to form a ground fog of ice particles. [P17487; 13° S, 147° W]

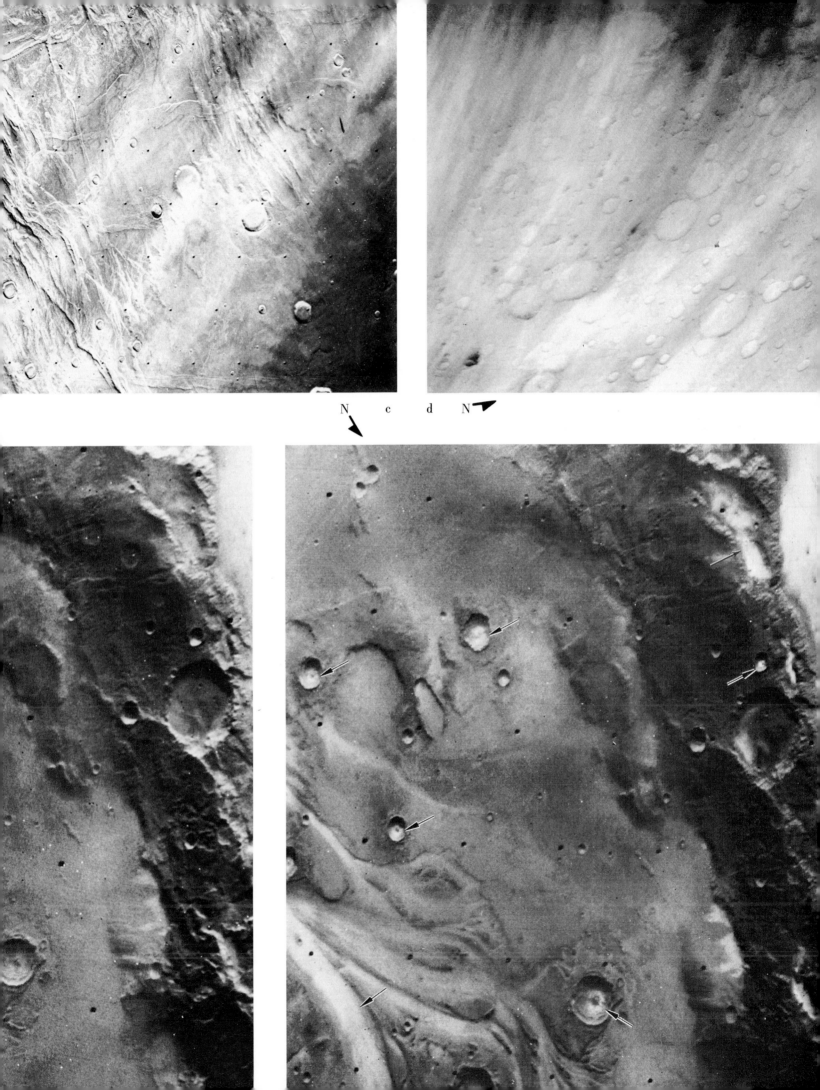

N c d N

Early Morning Clouds in Noctis Labyrinthus. Condensate clouds are seen here in early morning in the canyons of Labyrinthus Noctis, which lies at the western end of the equatorial Valles Marineris system. This picture, which covers about 90 000 km², was made by combining three frames of the same field taken through violet, green, and red filters. Although these clouds lie mainly down inside the canyons, they evidently extend above the walls and spill over some of the surrounding plateau. Like most condensate clouds in the Martian troposphere, they are believed to be composed of water-ice crystals. [P18114; 9° S, 95° W]

Dust Storm in Argyre Basin. A local dust storm in the Argyre basin near the end of winter in the southern hemisphere is seen from a relatively high-altitude point in the elliptical orbit of Viking 2. Winds appear to be coming from the west. The turbulent brown dust cloud near the polar cap boundary is roughly 300 km across. This cloud did not develop into a global dust storm of the type that tends to occur a little later in the Martian year when Mars is nearer to the Sun. Part of the receding seasonal frost cap covers the lower half of this picture. It appears yellowed by dust in the Argyre basin, but whiter in the mountains (at bottom of picture) at the southern rim of the basin. [P18598B; 50° S, 40° W]

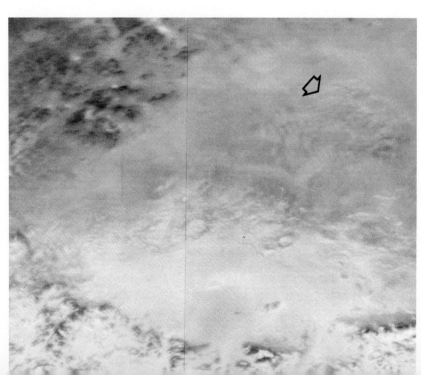

80 km

South Pole Dust Storm. This picture of the periphery of the retreating ice cap was taken the day after perihelion (Mars closest approach to the Sun). The cap had shrunk considerably since the time of the Argyre storm observation. The dust storm at the edge of the frost-covered area, which is just visible in the corner of the picture, is about 200 km across. Plumes of dust can be seen outside the boundaries of the main storm. This picture shows the first global storm in its last phase. Such storms are probably related to winds induced by great surface temperature contrasts. [248B57; 70° S, 60° W]

Local Dust Storms near Noctis Labyrinthus. The region southeast of the Noctis Labyrinthus complex on the slopes of the Tharsis bulge seems to be particularly conducive to the formation of local dust clouds. These frames were taken in the middle of spring (a) and in late spring (b). Both local dust storms occurred in the period between the two global dust storms. The area in which the local storms occurred slopes upward toward Arsia Mons. Infrared Thermal Mapper instrument data have shown that because of local differences in surface thermal properties, large temperature contrasts occur in this region. Downslope winds caused by these temperature gradients may be strong enough to create such clouds. [(a) 275B05-10, (b) 211B24; 14° S, 90° W]

400 km

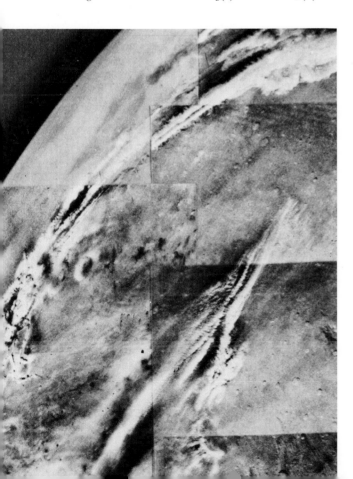

a

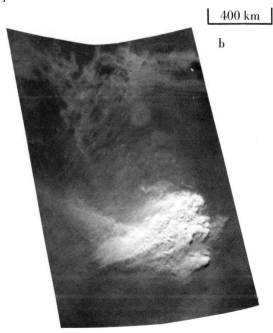

b

Dust Storm over the Chryse Basin. These two pictures of a dust storm over the Viking lander site in the Chryse basin were taken 170 seconds apart. Motion of the clouds can be detected if the pictures are viewed through a stereo viewer. Analysis of the two pictures indicates that portions of the cloud were moving from west to east with speeds ranging from 40 to 60 meters per second. This is consistent with westerly winds at the surface with the unusually high speed of 22 meters per second as recorded by the lander. The lander observation is, however, possibly in error because its wind sensor was damaged. [467A69, 467A31; 22° N, 48° W]

▷

Global Dust Storm. The early stages in the development of the first global dust storm were observed by Viking Orbiter 2 in the Thaumasia region. These images were taken 2 days apart. In (a), a single frame, imaged in red light from a very high altitude, includes the entire weather disturbance; the rest of the southern hemisphere was rather clear at this time. In (b), a mosaic, the frames were also taken through the red filter. They show an area several thousand kilometers wide seething with turbulent clouds of dust. This storm spread rapidly to higher altitudes, and suspended dust obscured much of the planet for a period of 50 days. Increased solar heating as Mars nears perihelion is thought to provide the energy that creates these large-scale disturbances. [(a) 176B02; (b) 211-5379; 42° S, 108° W]

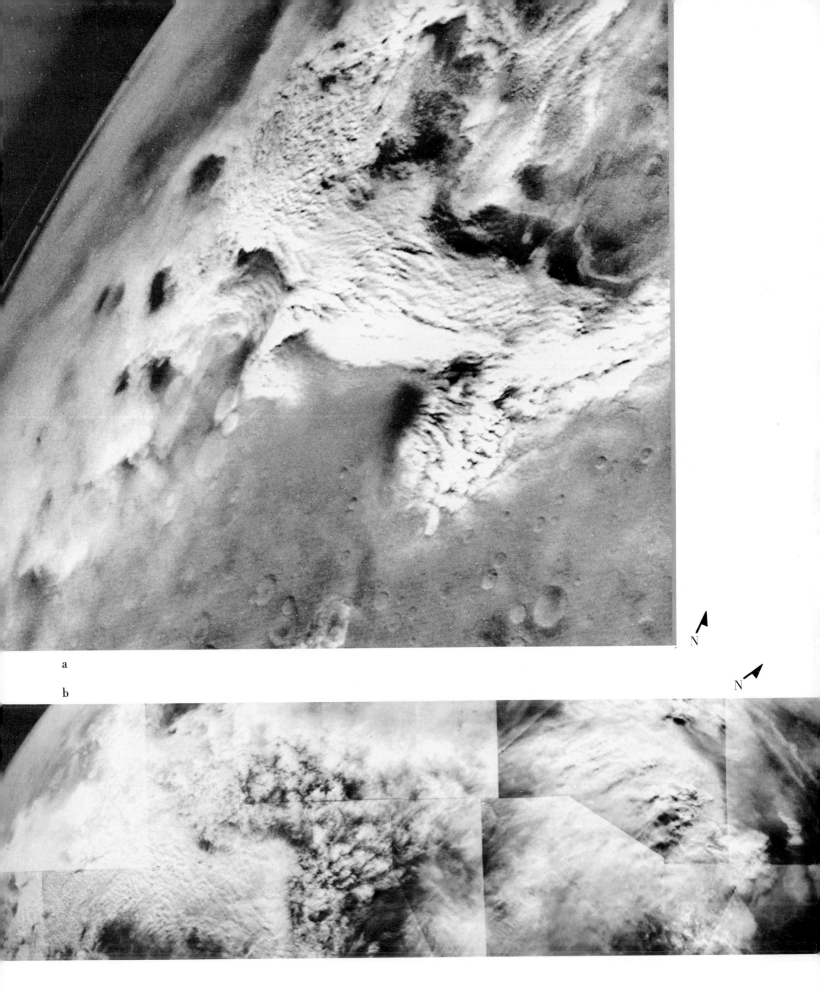

a

b

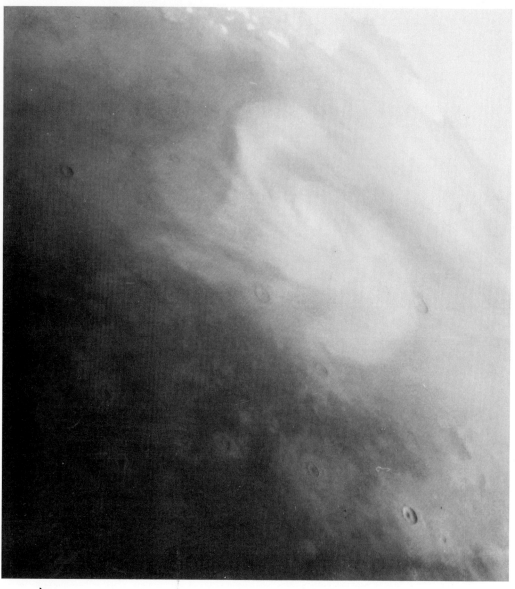

N

Low Pressure Cell near North Pole. This Martian storm was observed by Viking Orbiter 1 at about 65° N latitude. The local season corresponds to late July on Earth. The storm is located near Mars' polar front, a strong thermal boundary that separates cold air over the pole from the more temperate air to the south. Shadows indicate that the clouds are relatively low in the atmosphere. Because temperatures in this region are well above the condensation temperature of carbon dioxide, water ice is the probable constituent of the clouds. Water vapor concentrations are high (by Martian standards) during this season in the north polar region.

This system strongly resembles satellite pictures of extratropical cyclones near the polar front on Earth. The counterclockwise circulation is consistent with the winds expected in a normal low pressure situation.

The frost-filled crater Korolev (approximately 92 km in diameter) is located to the northeast of the storm. The white patches in the top center of the picture are outliers of the north polar remnant cap. [783A42; 70° N, 200° W]

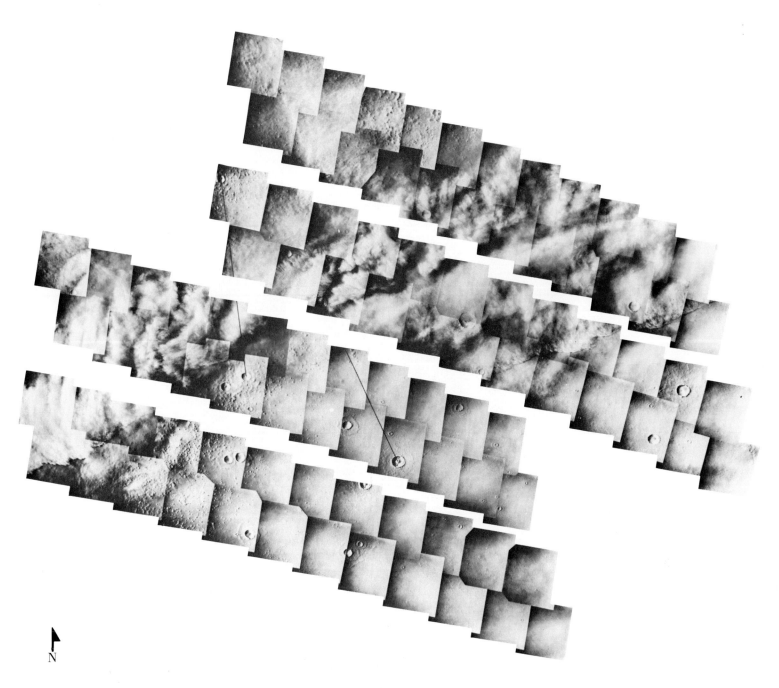

Cold Front. Viking Orbiter 2 photographed this cold front in the Arctic region when the season on Mars was equivalent to mid-May on Earth. During the 2 days between the upper and lower mosaics, the front moved about 950 km, at an average speed of 20 km/hr. The movement may be seen by comparing the two mosaics: lines connect identical features in the two sets of pictures. Weather systems like this appear to be common in Mars' northern hemisphere. Viking Lander 2 has detected the passage of similar fronts many times. Warm, comparatively moist air is lifted over a wedge of colder, denser air as it pushed south. Moisture in the warm air condenses into ice crystals, forming clouds. Dust, seen frequently in Martian storms, could also be present in these clouds. Some water must be present, scientists say, because wave clouds seen in both mosaics result from condensation and evaporation of ice crystals in the troughs and crests of pressure waves propagating in the atmosphere. [211-5764; 65° N, 135° W]

THE VIKING LANDING SITES

LANDING SITES for the Viking landers were chosen before launch, but close inspection from orbit showed them to be much too rugged and dangerous to risk landing on.

Viking Orbiter 1 searched for almost 3 weeks before finding a new, more suitable landing site about 800 km west-northwest of the original site, but still within Chryse Planitia. Because of the unexpected search for a new landing site, the original goal of landing on the historic date of July 4, 1976 was lost; the landing was achieved instead on July 20.

Viking Orbiter 1 took pictures of the proposed landing site for the second lander; from these it was determined, even before the second spacecraft arrived at Mars, that the site was not acceptable. Almost half of the planet's surface between 40° and 50° N was photographed in the attempt to find a suitable landing site for the second lander. Finally a site was found in Utopia Planitia, and the second successful landing was made on September 3, 1976.

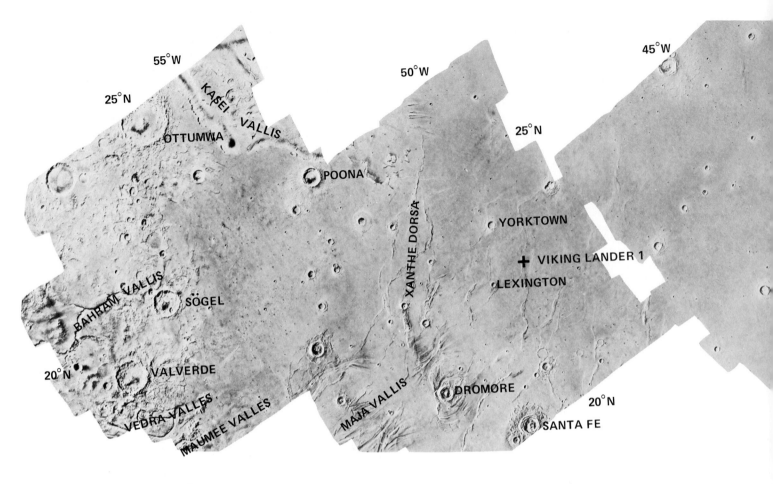

Chryse Planitia. This map is constructed from a series of images acquired by Viking Orbiter 1 on revolutions 10, 20, and 22. Chryse Planitia is an area of plains that seems to be volcanic in origin. The surface is characterized by mare-like ridges and relatively young impact craters, although subdued or part-filled craters are also present. At the left is a higher region of cratered terrain in which are several large channels. [USGS map M1M 23/50 CMC, 1977; I-1068]

High Resolution Mosaic of the Viking Lander 1 Site. The periapsis of Viking Orbiter 2 was lowered to 300 km to obtain high resolution images of the surface. The images here were taken by the spacecraft on revolution 452, and craters as small as 30 meters across can be seen. Craters A, B, C, and D are the same craters as shown in the map of the Yorktown region. Rims of craters C and E can be seen in images taken on the surface by Viking Lander 1. [452B09-11]

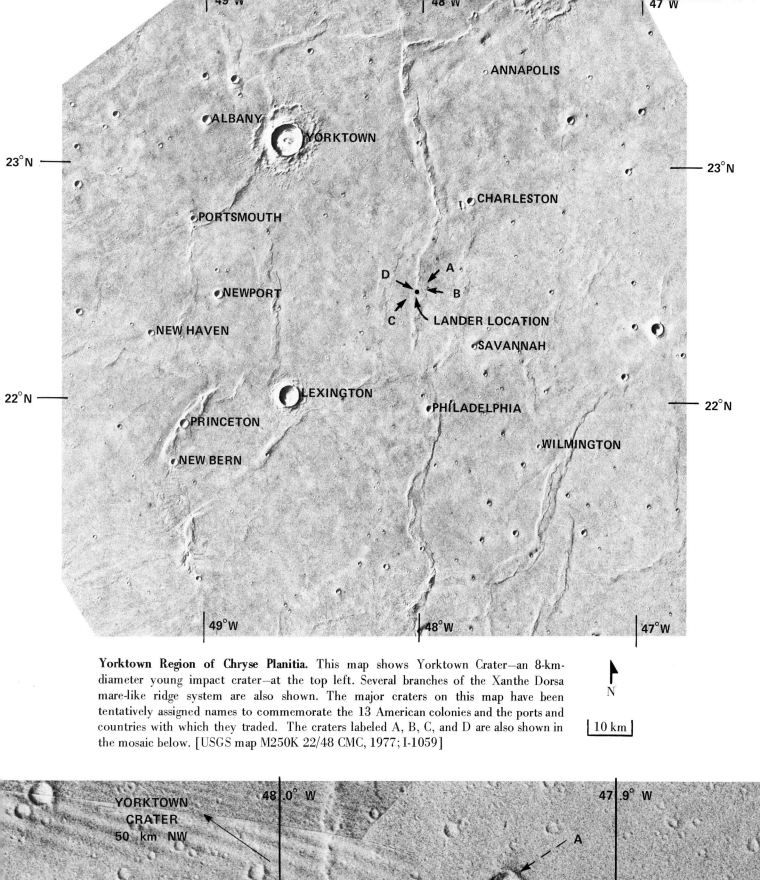

Yorktown Region of Chryse Planitia. This map shows Yorktown Crater—an 8-km-diameter young impact crater—at the top left. Several branches of the Xanthe Dorsa mare-like ridge system are also shown. The major craters on this map have been tentatively assigned names to commemorate the 13 American colonies and the ports and countries with which they traded. The craters labeled A, B, C, and D are also shown in the mosaic below. [USGS map M250K 22/48 CMC, 1977; I-1059]

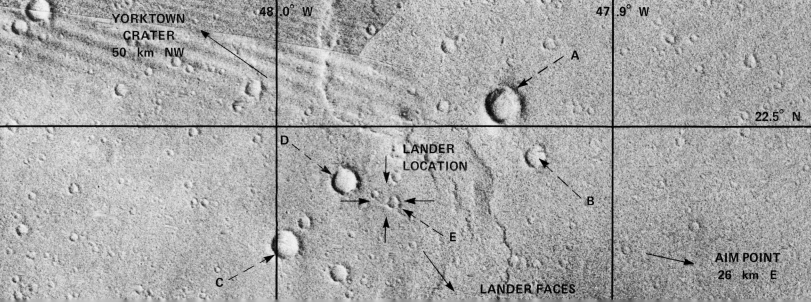

Panorama of Landscape. This view of part of the horizon around Viking Lander 1 was taken shortly after landing. At the left, about 100 meters away, is the northwest portion of the rim of Crater E in the high resolution mosaic on the previous page. At the right is the rim of Crater C, about 1.8 km away. The slight depression in the foreground just to the right of center may be a shallow, 3-meter-diameter secondary impact crater. [Viking Lander 1 Camera Event 12A002]

Mosaic of the Utopia Planitia Region. Viking Orbiter 2 acquired these pictures on revolution 9 during landing site selection and certification for Viking Lander 2. This part of Utopia Planitia is a polygonally fractured surface covered by a thin layer of presumably windblown material that has preferentially accumulated in the fractures. At the extreme upper right is Mie, a 100-km-diameter impact crater, approximately 200 km east-northeast of the lander. Mie Crater is covered by sand dunes and deflation hollows, both of which are strong evidence of wind activity. To the right of bottom center is Hrad Vallis. [USGS map M1M 46/230 CM, 1977; I-1061]

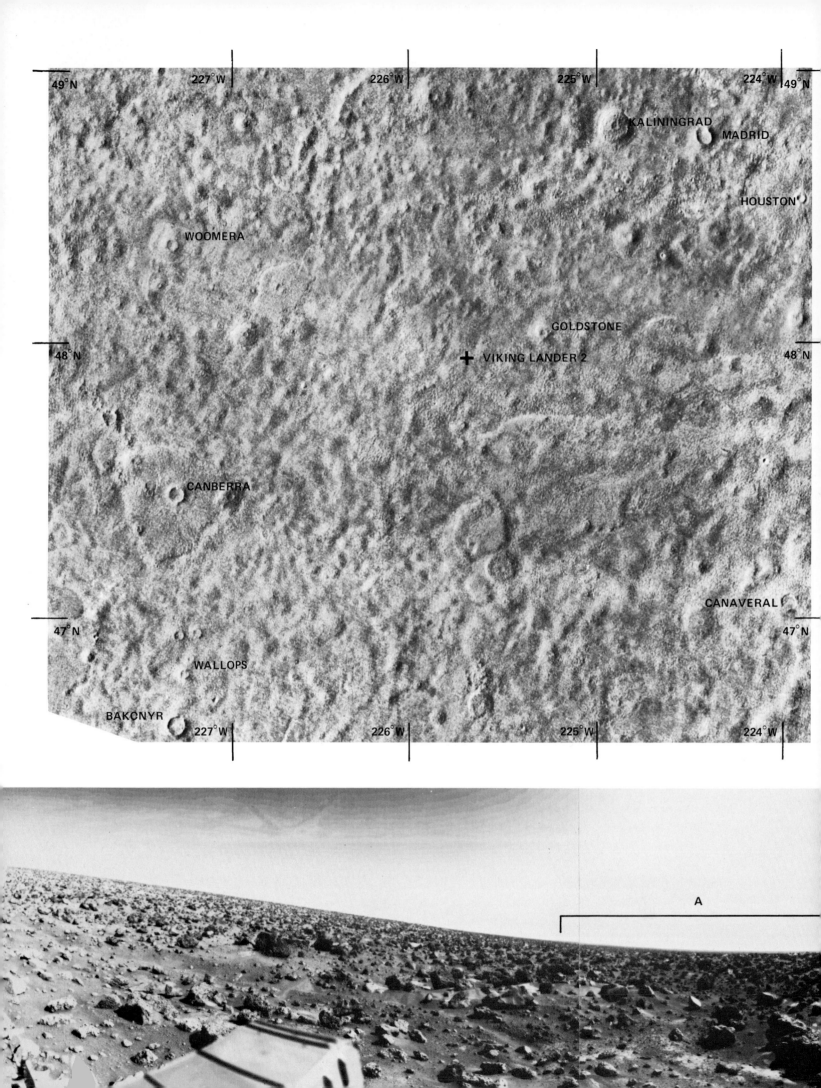

Mosaic of the Canberra Region. This mosaic, an enlargement of part of the mosaic of the Utopia Planitia region on the previous page, shows the detail around the site of Viking Lander 2. Most of the craters in this region are the pedestal type, produced when wind scoured away the softer, more erodable material between craters and left exposed the harder, more resistant material around the crater rims. The larger craters have been tentatively named for launch facilities, tracking stations, and mission control centers used in 1976. [USGS map M250K 48/226 CM, 1977; I-1060]

◁

N

| 10 km |

Panorama of Landscape. This panorama of the terrain surrounding the site of Viking Lander 2 was taken by the spacecraft lander shortly after touchdown. The large number of rocks and boulders was a surprise because prelanding evidence suggested that the area would be covered by a thin (tens of meters) sand sheet. Feature A on the far horizon is believed to be ejecta from the rim of Mie Crater. The large dip in the horizon was caused by the lander being tilted. [Viking Lander 2 Camera Event 22A002]

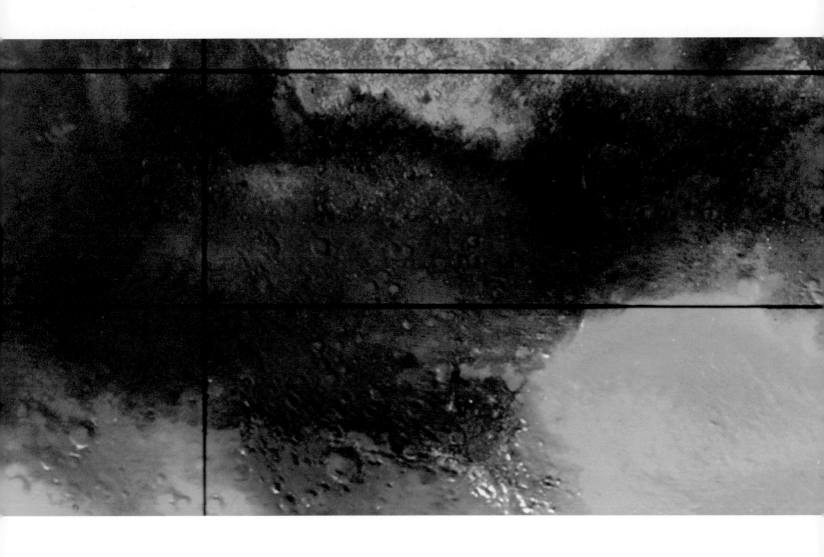

GLOBAL COLOR

DURING THE FINAL APPROACH to Mars, just before going into orbit, the Viking orbiters took a series of pictures of the planet with different color filters. As the planet rotated, different parts were photographed until almost the entire surface was covered. The pictures taken with the various filters were then compared, and the color at each point on the surface was determined by means of a variety of sophisticated computer processing techniques. The results are displayed here as Mercator maps of the equatorial regions. The approach trajectory prevented good visibility of the polar regions, and they are not shown.

Although Mars has very distinct light and dark areas, differences in color are quite subtle. The appearance of the planet is dominated by variations in brightness or albedo that cause the classical markings known from telescopic observations. The color of the surface is a fairly uniform rusty brown, and differences are so small that they can barely be seen. To see the color differences, they must be artificially enhanced through computer processing; at the same time, differences in brightness must be suppressed. Two color maps are shown here, one that approximates the natural color and another in which the color differences are artificially emphasized.

Color is of considerable geologic interest because it allows remote detection of chemical and mineralogical differences. Only the upper few millimeters of the surface contribute to the color, and on Mars this layer may be mostly wind blown debris. The bright materials that dominate the north equatorial zone are apparently aeolian deposits. Two units have been recognized. The upper unit is discontinuous, very red, and among the brightest of materials exposed at the planet's surface. The lower unit is darker and less red. The boundary between the two is generally serrated and has no relief. In the southern equatorial belt, the color variations are apparently related to local bedrock and not to randomly dispersed aeolian debris. The dark highland region (0° to 40° S and 60° W to 30° E) is divided into (a) dark red ancient crater rims, rugged plateaus, mostly riddled with small channels, and graben; and (b) dark "blue" volcanic flows intermediate in age, and show very few channel networks. The large volcanic constructs in the Tharsis region and volcanic centers in the southern highlands northeast of Hellas are both very dark and very red.

Atmospheric phenomena and surface frost affect the planet's appearance. South of approximately 40° S, the scene is dominated by the annual south polar carbon dioxide ice cap. Near-surface condensate clouds are abundant in this region, especially in Hellas. Because some of the data in the bright areas were saturated, the color balance is distorted; no attempt was made to correct this problem. North of about 20° N, condensate clouds are especially noticeable along the northernmost edge where emission angles were extreme. Other clouds are scattered locally throughout the equatorial region south and southwest of Valles Marineris.

Enhanced Color of Mars. In this image, all three color components have received the same contrast enhancement, which approaches saturation in the brightest areas. Because Mars is by a factor of two to three more reflective in the red than in the violet, the red component is predominant—giving the planet its classic rusty appearance. Some artifacts of the processing remain in the image, for example, diagonal streaks running from upper left to lower right.

Color Variations of Mars. This image dramatically enhances subtle color variations. The violet/green ratio is used as the blue component of the final image, the albedo at the green wavelength as the green component, and the red/green ratio as the red component. Hence, the amount of red or blue is controlled primarily by the slope of the spectral reflectance curve; areas with high albedo are also green. Thus, high albedo blue areas (ice, fog, clouds) are blue-green in color, and high albedo red areas are orange and yellow; bright areas of average color are green. Green is absent in dark areas, so the colors represent the slope from violet to red; red areas have a steeper slope, increasing from violet to blue; blue areas have a shallower slope.

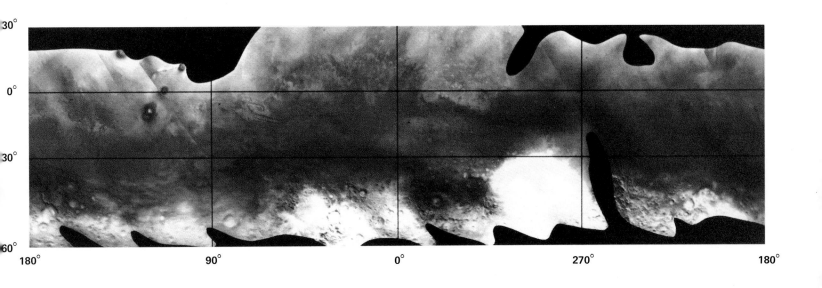

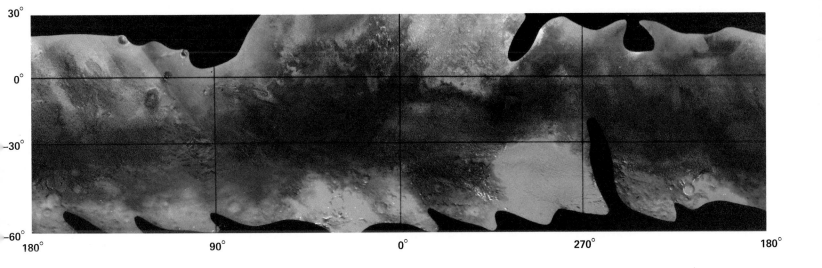

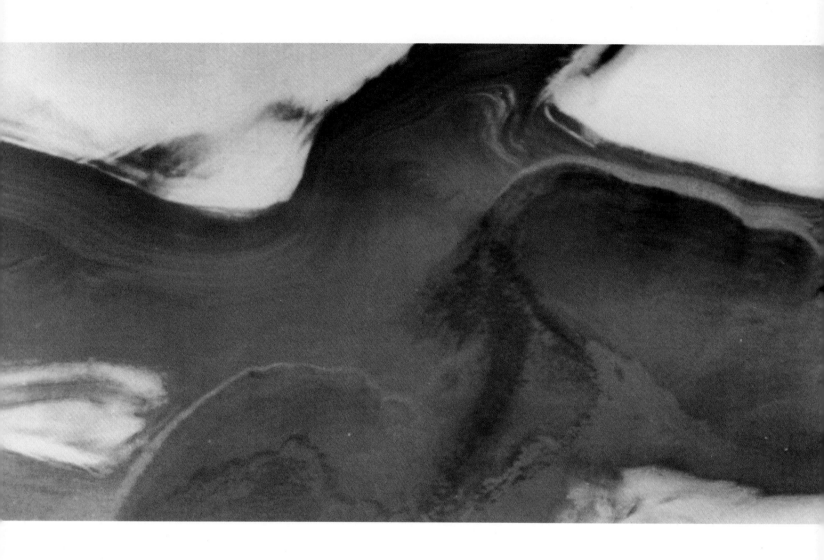

APPENDIXES

APPENDIX I
GLOSSARY

Aeolian.—A term applied to wind erosion or deposition of surface materials.

Albedo.—The reflectivity of a body compared with that of a perfectly diffusing surface at the same distance from the Sun, and normal to the incident radiation.

Apoapsis.—That point in an orbit farthest from the center of attraction.

Barchan.—A moving, isolated, crescent-shaped dune. The convex surface points toward the wind.

Basalt.—A dark, fine-grained volcanic rock. Very common.

Bench.—A small terrace or step-like ledge breaking the continuity of a slope.

Caldera.—A large volcanic depression containing volcanic vents.

Catena.—Crater Chain. A chain or line of craters.

Cavitation.—Plucking of material from the floor of a channel caused by the sharply reduced pressures associated with extreme flow velocities.

Chaotic terrain.—A surface consisting of short, jumbled ridges and valleys.

Chasma.—Canyon. An elongated, steep-sided depression.

Chondrite.—A stony meteorite characterized by chondrules embedded in a finely crystalline matrix consisting of orthopyroxene, olivine, and nickel-iron, with or without glass.

Collapse pit.—A closed, rimless depression caused by subsidence.

Dike.—A near vertical, planar, volcanic intrusion.

Dorsum (Dorsa).—Ridge(s). Irregular, elongate prominence.

Ejecta.—Material thrown out of an impact crater during formation. Such material may be distributed around a crater in distinctive patterns forming "ejecta rays" or "ejecta loops."

Escarpment.—A long, more or less continuous cliff or relatively steep slope produced by erosion or faulting. See "scarp."

Esker.—A long, low, narrow, sinuous, steep-sided ridge or mound composed of irregularly stratified sand and gravel that was deposited by a subglacial or englacial stream flowing between ice walls or in an ice tunnel of a continuously retreating glacier, left behind when the ice melted.

Etch pit.—A surface depression caused by the preferential removal of less resistant material.

Fault.—A surface or zone of rock fracture along which there has been displacement, from a few centimeters to a few kilometers in scale.

Folding.—The curving or bending of a planar structure such as rock strata, foliation, or cleavage by deformation.

Fossa (Fossae).—Ditches. Long, narrow, shallow depression. They generally occur in groups and are straight or curved.

Fretted.—Eroded in such a manner as to produce two horizontal planar surfaces separated by near vertical escarpments.

Fretted channel.—Long, relatively wide, flat floored valley with tributaries. Mass wasting probably played a significant role in their formation.

Gelifluction.—Creep of frozen material.

Glacioaeolian.—Material removed from a glacier by wind erosion.

Graben.—An elongate, relatively depressed crustal unit or block that is bounded by faults on its long sides.

Gradation.—The leveling of the land, or the bringing of a land surface or area to a uniform or nearly uniform grade or slope through erosion, transportation, and deposition.

Inclination.—The angle between the plane of an orbit and a reference plane. The Mars equator is here used as the reference when referring to spacecraft inclination.

Interfluve.—Lying between streams.

Labyrinthus.—Valley complex. Complex, intersecting valleys.

Laminated terrain.—A surface made of layers of different types of materials: layered terrain.

Lava.—Rock from a volcano, generally molten when ejected.

Limb.—The outer edge of a planetary disk.

Lithosphere.—The solid outer portion of a planet.

Mare.—Low-lying, level, relatively smooth, plains-like areas of considerable extent.

Mass wasting.—A term that includes all processing by which soil and rock materials fail and are transported downslope predominantly en masse by the direct application of gravitational body stresses.

Mensa (Mensae).—Mesas. Flat topped prominence with cliff-like edges.

Mons (Montes).—Mountains. A large topographic prominence or chain of elevations.

Normal fault.—A break in the surface caused by tensional forces.

Orographic.—Pertaining to mountains, especially in regard to their location and distribution.

Outflow channel.—A large-scale channel that starts at full width in chaotic terrain and has few, if any, tributaries.

Overthrusting.—A low-angle thrust fault of large scale, generally measured in kilometers.

Pahoehoe.—A type of lava having a glassy, smooth, and billowy or undulating surface; it is characteristic of Hawaiian lava.

Patera.—Irregular crater or a complex one with scalloped edges.

Pedestal crater.—A crater around which less resistant material has been removed from the ejecta leaving an elevated surface of more resistant ejecta material.

Periapsis.—The orbital point nearest the center of attraction.

Periglacial.—Said of the processes, conditions, areas, climates, and topographic features at the immediate margins of former and existing glaciers and ice sheets, and influenced by the cold temperature of the ice.

Perihelion.—That point in the orbit of a planet when it is closest to the Sun.

Phase angle.—The angle between a line from the Sun to the center of a body and a line from the spacecraft to the center of the same body.

Planitia.—Plain. Smooth low area.

Planum.—Plateau. Smooth elevated area.

Pleistocene.—A recent geologic epoch of the Quaternary period beginning approximately one million years ago, the last glacial age.

Polygonal ground.—Patterned ground marked by polygon-like arrangements of rock or soil. Generally produced on Earth by ice-wedge polygons.

Precessing ellipse.—An ellipse in which the pole is changing direction.

Rampart.—A narrow, wall-like ridge.

Regolith.—A general term for loose material overlying bedrock.

Reverse fault.—See thrust fault.

Rift.—A narrow cleft, fissure, or other opening in rock (as in limestone), made by cracking or splitting.

Rille.—Relatively long, trench-like valley; has relatively steep walls and usually flat bottoms.

Runoff channel.—Relatively small channel probably caused by water erosion over a long period.

Scabland.—Elevated, essentially flat basalt-covered land with only a thin soil cover.

Scarp.—A line of cliffs produced by faulting or by erosion. The term is an abbreviated form of escarpment, and the two terms commonly have the same meaning, although "scarp" is more often applied to cliffs formed by faulting.

Sediment.—Solid, fragmental material or mass of such material originating from the weathering of rocks, e.g., sand, gravel, mud, alluvium.

Shield volcano.—A broad, gently sloping volcano.

Striae.—Striped ground.

Subduction zone.—An elongate region in which a crustal mass descends below another crustal mass.

Subsidence.—A localized gradual downward settling or sinking of a surface with little or no horizontal movement.

Tectonic.—A term pertaining to deformation of a planet's crust, especially the rock structure and surface forms that result.

Terminator.—An imaginary, diffuse line separating the illuminated and dark portions of a celestial body. There are two terminators: morning and evening.

Thermokarst.—Rimless depressions caused by the melting of ice and subsequent collapse of the surface.

Tholus.—Hill. Isolated domical small mountain or hill.

Thrust fault.—A fault caused by compressional forces.

Transcurrent faulting.—A large-scale strike-slip fault in which the fault surface is steeply inclined.

Troposphere.—The lowest layer in an atmosphere, generally considered to be 10-20 km thick.

Tuff.—Volcanic ash, particles of 4 mm diameter or smaller.

UTC.—Universal Time Coordinated.

Unconformity.—The relationship where the younger upper strata do not follow the dip and strike of older underlying strata.

Vallis (Valles).—Valley. A sinuous channel, many with tributaries. These are named "Mars" in many languages, e.g., Al Qahira Vallis is derived from the Arabic word for Mars.

Vastitas.—Extensive plain.

Yardang.—Elongated, sculpted ridge formed by wind erosion.

SOURCES

American Geological Institute: *Glossary of Geology, 1972.*

Robert J. Foster: *Physical Geology.* Charles E. Merrill Publishing Co., Columbus, Ohio, 1971.

Richard M. Pearl: *Geology.* Barnes & Noble, Inc., New York, NY, 1969.

G. DeVaucouleurs, et al.: "The New Martian Nomenclature of the International Astronomical Union," Icarus 26, 85-98, 1975.

APPENDIX II
THE VIKING ORBITER IMAGING SYSTEMS

THE VIKING ORBITER cameras evolved from the cameras flown on the earlier Mariner spacecraft. Each generation of spacecraft sent to Mars has featured cameras with vastly improved capabilities, especially higher resolution and increased light sensitivity. Basically the cameras are high-performance vidicons, similar to those used in television cameras on Earth, and have a telephoto lens assembly in front of them. Each orbiter has two identical cameras. The cameras, along with instruments to measure the surface temperature and amount of water vapor in the atmosphere, are mounted on the science platform, a device that can be moved about two axes to achieve a scanning motion when viewing the Martian surface from orbit. The drawing shows the arrangement of instruments on the science platform.

Each camera, along with its 475-mm focal length telephoto lens, has a field of view of 1.54° × 1.69°. From an orbital altitude of 1500 km, each frame covers a minimum area on the surface of 40 × 44 km. Six selectable filters allow color images to be acquired. Acquisition of a frame and subsequent readout to the tape recorder requires 8.96 seconds, so a frame is taken on alternate cameras every 4.48 seconds. This alternating pattern, coupled with motion along the orbit, combines to produce a swath of pictures, as shown in the illustration of orbiter imagery coverage.

The image on the vidicon is scanned, or read out, as 1056 horizontal lines. Each line, in turn, is divided into 1182 pixels (picture elements), and the brightness of each pixel can range from 0 to 127 arbitrary units. Thus, to record a single frame requires the storage of almost 10 million bits (binary digits) on the orbiter tape recorder. The pictures are stored on the tape recorder in digital form until there is an opportunity to play back the data over the orbiter communications system to a receiving station on Earth.

As the data are received on Earth they are subject to computer-based image processing. All images receive first-order processing that consists of the following: noise removal, contrast enhancement, and shading correction. First-order images are the most widely used for scientific analysis and are the version most commonly seen in this book. Some images also undergo a more complex second-order processing that includes all of the first-order processing plus sophisticated procedures to remove geometric and radiometric distortions and merge black and white images taken through different wavelength filters into a composite color image. Second-order processing can also include techniques such as generating stereoscopic pairs and picture differencing in which an image taken at one time is digitally subtracted from another image of the same scene at a different time to highlight any changes that may have occurred during the interval between the images.

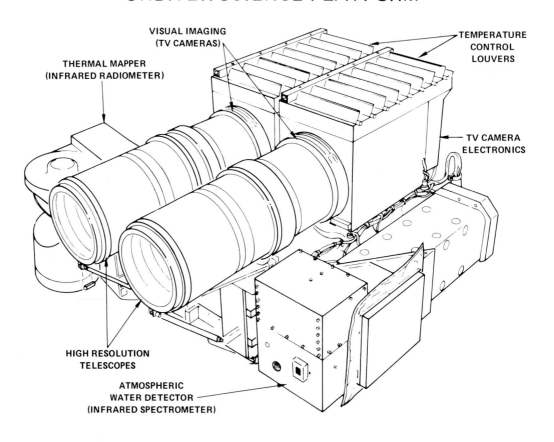

ORBITER SCIENCE PLATFORM

APPENDIX III
OTHER SOURCES OF VIKING DATA

THE RESULTS OF PROJECT VIKING have appeared in countless popular and scientific publications, which it would be futile to list. However, the reports given below represent the best summary of the scientific findings:

>Journal of Geophysical Research
>Volume 82, Number 23
>September 30, 1977
>Volume 84, Number B6
>June 10, 1979
>Volume 84, Number B14
>December 30, 1979
>
>*Published by:*
>American Geophysical Union
>1009 K Street, N.W.
>Washington, DC 20006

A comprehensive bibliography of major Project Viking papers, both scientific and engineering, can be ordered from:

>Bibliography of Viking Mars Science
>Mail Code 111-100
>Jet Propulsion Laboratory
>4800 Oak Grove Drive
>Pasadena, CA 91103

A summary of how the total Viking System works can be found in:

>Spitzer, C. R.: "The Vikings Are Coming," *IEEE Spectrum*, Vol. 6, June 1976, p. 48-54
>
>*Published by:*
>Institute of Electrical and Electronics Engineers
>345 E. 47th Street
>New York, NY 10017

Qualified scientists and educators may request Project Viking scientific data from:

National Space Science Data Center
Code 601.4
Goddard Space Flight Center
Greenbelt, MD 20771

For the student of Mars interested in examining the companion to this volume, which displays selected lander photographs, please see:

The Martian Landscape
NASA SP-425
National Aeronautics and Space Administration
Washington, DC 20546

For sale by the Superintendent of Documents
U.S. Government Printing Office, Washington, D.C. 20402.
(Stock number 033-000-00780-9)

APPENDIX IV
PROJECT VIKING MANAGEMENT PERSONNEL

Prelaunch, Cruise, and Primary Mission

    Walter Jakobowski, Program Manager
    Richard S. Young, Program Scientist
    James S. Martin, Jr., Project Manager
    A. Thomas Young, Mission Director
    Gerald A. Soffen, Project Scientist

Extended Mission

    Walter Jakobowski, Program Manager
    Richard S. Young, Program Scientist
    G. Calvin Broome, Project Manager-Mission Director
    Conway W. Snyder, Project Scientist

Continuation Mission

    Guenter K. Strobel, Program Manager
    Joseph M. Boyce, Program Scientist
    Kermit S. Watkins, Project Manager-Mission Director
    Conway W. Snyder, Project Scientist